A PHILOSOPHICAL REJECTION
OF

THE BIG BANG THEORY

By Khuram Rafique

Copyright © 2018 by Khuram Rafique
All rights reserved. This book or any portion thereof
may not be reproduced or used in any manner whatsoever
without the express written permission of the publisher
except for the use of brief quotations in a book review.

ISBN-13:
978-1986907378

ISBN-10:
1986907376

Ebook ASIN: ASIN: B079R8PM4X

Contents

THE BIG BANG THEORY 1
I. FOUNDATION OF THE BIG BANG MODEL 1
II. OBSERVATIONAL SUPPORT 39

Preface

The analysis in this book is started with the confirmed fact that Alexander Friedmann's 1922 work had no relation with Hubble's Law that was yet to be found by Edwin Hubble in 1929. Official sources repeatedly tell us that Georges Lemaître had found similar to Friedmann's solution in year 1927 so I thought that Lemaître's work also should have no actual relation with Hubble's Law. My analysis kept going with this assumption till section I.III where I realized that if unlike Friedmann, Lemaître had the data of Doppler's Redshifts of various galaxies then he also could have means to find the distance of those galaxies. Admittedly, this book up to section I.III is an analysis based on an incorrect assumption that by 1927, Lemaître should be unaware of Hubble Type redshift-distance relationship in light coming from far off galaxies. But that analysis forced me to download 1927 paper of Lemaître. Initially I found English Translation (1931) by the title: "A Homogeneous Universe of Constant Mass and Increasing Radius accounting for the Radial Velocity of Extra-galactic Nebulæ". I was shocked to see that my analysis was wrong up to section I.III because apparently Lemaître had already derived Hubble type redshift-distance relationship solely from General Relativity (GR) Equations. But I was not wrong. This was a manipulated translation; he had not derived that relationship from GR equations rather had derived from a method that he took from Hubble himself, detail thereof I have explained in this book. Here in this book, original papers of Alexander Friedmann (1922), Georges Lemaître (1927), Edwin Hubble (1929), Albert Einstein (1917) along with other important relevant papers have been analyzed and only the most fundamental aspects like expansion and CMBR of the Big Bang Cosmology are covered. If these two aspects of the Big Bang Cosmology are precisely refuted then there is nothing crucial left with the standard model.

Philosophy is not concerned with providing definite solutions to the problems. Therefore, alternatives suggested in this book should not literally be taken as definite alternatives. They however represent philosophically solid and justified positions and it is up to readers who should conclude the matter by applying their own critical judgment. This book will however expose the undue authoritative nature of FLRW metric and with this book, The Big Bang Theory is set to become a story of past.

Khuram Rafique (2018)
Book's Blog: https://bbtrejected.wordpress.com

✳ ✳ ✳

I. FOUNDATION OF THE BIG BANG MODEL

I.I. A Philosophical Review of the Big Bang Theory is warranted

Twentieth century had been remarkable with regards to scientific and technical developments. Real scientific progress converted to technological advancements that resulted in a paradigm shift in human way of living. Nevertheless, towering intellectual achievements of twentieth century are not unquestionable. No doubt science progressed – but so did huge bangs of intellectual fallacies. Highly educated people now keep on telling incomprehensible things as hard facts of science. A fashion of promoting 'counter intuitive' theories of Physics emerged. Metaphysics of Philosophy was discarded altogether but science itself assumed the shape of metaphysics. One such metaphysical theory of modern science is the famous Big Bang theory which is the subject matter of this book. This book is a philosophical review of the Big Bang Theory of Modern Physics.

In essence, science ought to be an understanding developed out of careful real observations or experiments but modern science has, to a great extent, replaced real observations with equations of mathematics. Mathematics vs. experimentation, logic or commonsense is not the topic of this book as this topic will be covered in my other book on Epistemology. Here I will only show that whole edifice of the Big Bang Theory rests only on single pillar of mathematics which is not supported by real observations. Given this fact, the Big Bang Theory should be regarded as a Philosophical or Metaphysical Theory rather than part of science. More to this, the theory is not even legitimate metaphysics as the theory dodges the reader into wrongfully believing that it is based on real observations. At the most, they have a mathematical model as foundation. Real observation is not the part of foundation. Then on the basis of a dubious mathematical model, real observations are explained to show that observed reality has been a possibility due to the reason that proposed mathematical model is accurate. The argument is that observed reality cannot be explained except with the help of mathematical model of Big Bang or at least that Big Bang is the best explanation that we have of observed phenomena. Philosophical review of the theory is needed because after all it is not based on real observations and the task of only explaining observed phenomenon can be handled by Philosophy as well. Therefore this book will not only show that the Big Bang Theory is misleading and unscientific, here an outline of alternative possible explanations of observed phenomena shall also be presented. However, this book is not going to offer definite alternative because to work out a definite and detailed as well as correct model is still the task of science that will be done after getting assurance that model is not without real observations as part of foundation. If foundation remains devoid of real observations and model itself comes at foundation level to only account for or interpret real observations then the model shall remain part of philosophy.

I.II. How the Big Bang Theory Dodges the Reader into believing that it is based on Real Observations?

There is definite dodge as I shall explain here shortly. But unlike malevolent fraud, it is more like an uncorrected mistake. It happened that mistake was duly realized – but mathematics was developed (or modified) to stay with the earlier incorrect understanding. It was all started with real observations when as early as year 1912, Scientists started noticing redshifts in far off galaxies (then thought of spiral nebulae as there was no concept of separate galaxies by that time). Naturally, those redshifts were interpreted in terms of Doppler's Effect. In 1922, Alexander Friedmann and then in 1927, Georges Lemaître had data of Doppler's Shifts and both of them formulated their equations depicting an expanding universe. Friedmann might not actually have employed Doppler's Shift data as he only derived mathematical models of expanding or oscillating Universe solely from available solutions of General Relativity equations. Lemaître however employed Doppler's Shift data in the formulation of his equations.

So far, overall approach was not unscientific because Friedmann's model was only abstract mathematics and he had not presented that model as a confirmed scientific fact. For the case of Lemaître, equations were derived out of available observational data relating to Doppler's Effect, in combination with the same solutions to GR equations wherefrom Friedmann already had derived his results. Apparently that was a scientific approach because along with mathematical solutions to available equations, observational data was also considered during the process of derivation of results. But the approach was scientific only in a superficial mode. In fact, there was no Doppler's Effect in redshifts coming from far off galaxies. Interpretation of redshifts in light coming from far off galaxies in terms of Doppler's Effect was an incorrect interpretation and initial 'scientific' theory of Big Bang in the form of equations of Lemaître (1927) was based on incorrect interpretation of observational data. Mathematics was not immune to interpretational errors of observed data. Mathematics itself was capable to formulate model of any kind of Universe whether it was expanding, contracting, static, pulsing, swirling, churning or whatever kind of Universe. But this type of abstract mathematics would be suitable if the task was to construct a whole new Universe from scratch. But within the domain of Physics, task of mathematics was only to construct a representative mathematical model of real physical Universe. To construct a representative model was not the task of Friedmann because he only provided abstract mathematics where he explored all the possibilities; ruled out few of them as impossible but acknowledged few other options (including expanding universe option) as mathematically possible. However goal of Lemaître was to construct a representative model of real world but he ended up with a representative mathematical model of a wrong interpretation of observed data. The task was to construct representative mathematical model of correct interpretation of reality where cause of redshifts was yet to be determined but achievement was a representative mathematical model of incorrect interpretation of reality where cause of a different kind of redshift was taken to be the same Doppler's type receding of objects. Off course, mathematics is capable to construct representative model of any interpretation; no matter right or wrong. But if it is a representative model of misinterpretation of reality then sooner or later reality itself will notify us an error message of 'mismatch'. There should, however, be someone having ability to read that error message. Nevertheless, fixing that error would require separate set of abilities. In this case, soon Edwin Hubble was going to read an error message because he was closely watching the reality at that time.

I.III. Doppler's Effect was Not the

reason of Redshifts in Light coming from far off Galaxies

In year 1929[1], Edwin Hubble first time noted that more distant an object was the more red-shifted was the light coming from that object. Its clear meaning was that scientists had not been observing Doppler's Shifts since the beginning of 20th century. Actual thing came out in 1929 was that it was a different kind of redshift which is called Cosmological Redshift.

Doppler's Shift (redshift) is observed if something is physically moving away from us. Let's say a far off galaxy is physically moving away from us. The light emitted by the galaxy, right from start, will be redshifted to the full value. With Doppler's Shift, we get a physical proof that yes the galaxy is physically moving away from us.

Whereas in 'Cosmological Redshift', the far off galaxy is physically not moving away from us and normal light is emitted by that galaxy. But during long journey of light, wavelength of light keeps on increasing. The larger distance is covered, the wavelength has become larger. It means if larger distance is covered, the greater redshift is observed at the receiving end. Exact this thing was noted by Edwin Hubble and finally scientists realized that what redshifts they had been observing since second decade of twentieth century were not Doppler's Shifts but were Cosmological Redshifts.

At this point, the Wikipedia article on Edwin Hubble states that "yet the reason for the redshift remained unclear".

One thing is however clear by now. Galactic redshifts had been interpreted in terms of Doppler's Effect up to the year 1929. By that time, reason for the redshift was clearly known to be the Doppler's Shift and clarity of this reason was not doubted. But the finding in 1929 that redshift increases with increased distance ruled out Doppler's Effect as the underlying reason for the redshift and the actual reason for the redshift became unclear. However, dominant science people promoted the idea that Hubble type redshift-distance relationship was predicted two years before by Lemaître. But Hubble never conceded to this promotion of Lemaître.[2] Hubble was the one who had successfully read the error message in reality relating to previous Doppler's based understanding of galactic redshifts and he remained skeptical to the whole idea of expansion. Allan Sandage informs us that, for Hubble, recession of galaxies was not the final meaning of redshifts as the redshifts could represent unrecognized principle of nature.[3]Hubble was a real scientist; he never fell towards expansionist regime though he also could not openly oppose them. The actual thing that we learnt in twentieth century was not that spacetime is curved or that Universe started with a Big Bang out of singularity or other like metaphysical things. The concrete scientific facts that we learnt in 20th century were that there are real island universes (galaxies) or that the more the distance of a galaxy from us, the more redshifted is the light that we receive from that galaxy and we have learnt these two hard scientific facts from Edwin Hubble.

Now Hubble did not find connection between Lemaître's work and his own finding. But mainstream Physicists insist that Lemaître had already predicted Hubble Type redshift-distance relationship on the basis of Einstein's equations of General Relativity. Although Wikipedia article on Edwin Hubble accepts that, to date, no remaining papers or verification exist where any link between Lemaître's work and Hubble's measurements could be found yet the same article also insists that it is reality that Lemaître had already predicted Hubble type redshift-distance relationship on the basis of Einstein's equations of General Relativity. This position sounds like a kind of undue love of equations. For example, how GR equations could give him idea of Hubble type relationship of redshifts-distance when gravity is all about attraction or at the most, 'curvature'? Apparently, GR equations could have nothing to do with Hubble type relationship of redshifts-distance until and unless so-called cosmological constant is a wild form of anti-gravity that is more than an inverse curvature and resembles a straight line repulsion system. Here it might be true that in 1927, Lemaître took cosmological constant as a form of 'pressure of radiation' (i.e. a form of straight line repulsion agent) but if it was really the case then why he later on abandoned this 'erroneous' idea?

"Lemaître conceived the static Einstein universe as a kind of pre-universe out of which the expansion had grown as a result of an instability. As a physical cause for the expansion he suggested the radiation pressure itself, due to its infinite accumulation in a closed static universe, but he did not develop this (erroneous) idea."[4]

Lemaître had suggested reason of expanding Universe to be the 'radiation pressure' in his famous 1927 article. "He did not develop this idea" means that after 1931, he moved to his different suggestion of 'Primeval Atom' or 'Cosmic Egg' where he would not bring the original idea of 'radiation pressure'. If he really had developed his expansion theory on the basis of 'radiation pressure' then he should not later on have abandoned this idea. This idea was therefore not the core component of his 1927 work that's why he did not develop or pursue this idea later on. It also means that 'radiation pressure' was an orphaned idea that was not backed by equations. Now if 'radiation pressure' or any such physical factor was not the core component of his 1927 paper then without such a core component, it was not possible to derive Hubble type redshift-distance relationship solely from equations.

Therefore, our conclusion is that expansionist regime of that time unduly assigned credit of Hubble type redshifts-distance relationship to relativity based equations developed by Lemaître and Friedmann. Universe was not expanding but expansionist regime was set to expansion due to multitude of factors that we shall explore in coming pages.

Expansionists did not revert to the idea of expansion even after knowing that redshifts in light coming from far off galaxies was not due to Doppler's effect. Given the fact that unlike redshift-speed relationship (Doppler's Redshift), the actually observed redshift-distance relationship (Cosmological Redshift) was not the physical proof of receding of those far off galaxies; they had lost the observational basis to expansionist regime. But they unduly started saying that Hubble type redshift-distance relationship was already explained in Lemaître (1927) equations. No one took pain in doing hard work of finding the actual reason of redshifts despite the fact that Hubble requested prominent scientists of his time to come and provide satisfactory theoretical interpretation of the redshift-distance relation.[5] After having read the error message in reality, Now Hubble was looking for the right person who could fix that error but now no one was listening to him on this point.

Since satisfactory explanation of redshifts was not coming from anywhere, expansionist regime acquired more potential to grow. But expansionist regime was also caught up in additional troubles. Immediate problem realized was that 'then why do we appear to be at center of universe?' That problem was though solved immediately but question arises is that if redshifts-distance relation was already addressed by Lemaître's or even Friedmann's equations then why did both of them not solve the associated problem of 'why do we appear to be at center?'

Friedmann and Lemaître could not solve this problem because it was not a problem arising out of their equations. Friedmann had, in abstract mathematical terms, only talked that at the starting time of creation (i.e. time = 0), radius of Universe must be zero and he also has used word 'point'. After 10 billion years that point, as per equations, would reach to a radius that could hold 5×10^{21} solar masses (as accepted or guess of his time). Clearly he has only talked about relationship of expansion rate of Universe with time or mass density. He has not talked about relationship of expansion rate with already achieved expansion. For example mass density of 5×10^{21} solar masses that was initially concentrated on single point (though Friedmann has not stated this thing in physical meanings) could expand the Universe up to certain radius (of our present Universe) in 10 billion years. Ok, we accept it for the sake of argument. But with this, mass density has been reduced from what it was at initial stage. With lesser mass density then before, now onwards our rate of expansion should be reduced. And this is directly opposite to the accepted meaning of Hubble's finding according to which the greater the radius achieved, the greater should be the further expansion rate. The relationship of speed of expansion with already acquired expansion is nowhere in the works of Friedmann. He was simply unaware of yet to be found 'facts' of relationship of 'recessional velocities' with distance. Furthermore, Lemaître, in his 1927 article, has clearly assumed universe with a definite radius. With a Universe of definite radius, the problem of why we appear to be at center could not be solved and neither did Lemaître actually attempt to solve this problem even if he knew the redshift-distance relation. In short, it is plain lie to say that Friedmann's and Lemaître's equations had already accounted for redshifts-distance relationship on the basis of GR equations at the time when this relation was not discovered by Hubble. Hubble also had not accepted this lie as he actively sought satisfactory explanation of that relationship from the prominent relevant scientists of his time. But the lie was going to be supported by a 'mathematical proof'. By 1935, Robertson and Walker presented mathematical proof that Friedmann and Lemaître's equations had worked out spatial homogeneous and isotropic universe.[6] Consequently Friedmann- Lemaître-Robertson-Walker (FLRW) metric was declared to be the only possible interpretation of Hubble type redshifts-distance relationship and chapter was closed for any alternative explanation of redshifts. Although few alternative proposals emerged like 'tired light' or other justifications but all were discarded or might be they really failed at certain physical tests. But wherever expansion model fails a physical practical test, we always get non-physical fudge factors like 'expansion of space' and other similar absurd things. The same facility is however not available to the alternative explanations that's why they can be discarded easily.

Anyhow, the picture emerged so far is that – Friedmann and Lemaître already developed equations for expanding universe and at least Lemaître did employ Doppler's Shift data for working out those equations. The time when Hubble experimentally noticed the redshift-distance relationship, theoreticians rightfully or wrongfully realized that same relationship was already described by the equations of both the brilliant mathematicians despite the fact that brilliant persons prior to 1929 must have employed available Doppler's Effect data which should not have given redshift-distance relation rather should have given redshift-speed relation. But equations of both brilliant persons were further authenticated by the fact that those equations were derived out of super solutions to supreme equations of GR by Einstein. Hubble himself however failed to see any convincing relation between Lemaître's equations and his own findings and he tried to invite relevant scientists to come and provide solid theoretical explanation.

If we base our expansion model on Doppler's shift data then all we can get is a relation of increasing (recessional) speed with increasing redshift value. Now primarily 'distance' becomes irrelevant within the meanings of Doppler's Effect type expansion. At near distance, receding speed is 100; at greater distance, the receding speed again shall be 100. Both near and far objects, given that recessional speed is same, shall give same value of redshift. In other words, there shall be no redshift-distance relationship. However what we are told by the expansionist regime is that at least Lemaître based his expansion model on Doppler's Shift data but (since he also incorporated GR equations) he successfully achieved Hubble type redshift-distance relationship in his model. But we also have seen earlier that GR equations should not have provided him any hint of redshift-distance relationship. Given that Doppler's Effect data and General Relativity Equations were incorporated in expansion model, the maximum possibility was that expansion could be proposed due to available data of Doppler type redshifts and even a rate of expansion also could be proposed, again based on same available data.

I.IV. In 1931, Lemaître Suppressed Crucial Facts by Publishing Manipulated Translation of his own 1927 Article

Parallel to the above narrated expectations, there were however surprising actual events. In 1927, Lemaître did present a redshift-distance relationship which is acknowledged by the mainstream science community of today but that was not acknowledged by Edwin Hubble himself. Yes, there is proportionality relationship between redshift and distance in the article titled "A Homogeneous Universe of Constant Mass and Increasing Radius accounting for the Radial Velocity of Extra-galactic Nebulæ" (English Translation: 1931)[7]. But why plain proportionality relation of redshift and distance could not satisfy Hubble? Pro-Lemaître sources[8] directly blame Hubble that he never read actual paper of Lemaître that's why he failed to appreciate the fact that Hubble type redshift-distance relation was already derived from equations by Lemaître.

Therefore it is important that we may analyze what actually Lemaître had proposed in year 1927. The source we have is the translation of 1927 article by Lemaître himself published in year 1931 and also the original French article published in 1927. My finding is that there is huge blunder in the translation of article that was published in year 1931. But before explaining the blunder of 1931 translation, let us first see what the case in favor of the Big Bang Theory is that exists today in year 2018. The whole case of the Big Bang Theory is that although exact redshift-distance relation was experimentally found by Hubble in year 1929, but at least Lemaître had already derived same relationship from a solution to relativity equations. Since equations rightly picked the underlying reality, therefore the only reason of redshift-distance relation that was found by Hubble was same equations. Therefore, we should forget that Cosmological Redshift (redshift-distance relation) is different from Doppler's Redshift (redshift-speed relation) or that Cosmological Redshift, unlike Doppler's Redshift, is not the physical proof of receding of anything. Since equations rightly described Cosmological Redshift and since same equations described an expanding universe, therefore there is no need of physical evidence that Cosmological Redshift is also the proof of receding of anything. Perhaps we can use both terms 'Cosmological Redshift' and 'Doppler's Redshift' interchangeably which is actually being done in official papers and textbooks even today. It is exact this interchangeability of these two separate terms in official papers and science discussions that I call dodge that portrays Big Bang Theory as fully backed by experimental proof of 'Doppler's Effect'. In reality, we only have physical proof that yes Doppler's Redshift actually indicates receding of anything but we do not have any experimental proof that Cosmological Redshift is also proof of receding of anything. We have only mathematics.

Anyhow, the whole case of the Big Bang Theory rests on a single fact that relativistic equations (Lemaître's) predicted same Hubble type redshift-distance relationship almost two years before the actual experimental discovery of same relation by Hubble. But this single fact is a huge blunder. Yes 1927 French article had already 'discovered' that relationship (in less accurate form) but actually that relationship was not derived out of any equation. We have already seen in our previous analysis that given only the Doppler's Redshifts data and relativistic equations, Hubble type redshift-distance relationship could not be derived. But after having written this analysis when I actually downloaded the 1931 English translation of article, I literally remained astonished and dumbfounded to see that almost same Hubble type redshift-distance relationship was already contained in that article. But before I also fell into believing the magic of mathematics, I read in another pro-Lemaître paper[9] that there were certain discrepancies in original French article of 1927 and English Translation of 1931. Being pro- Lemaître, this paper at first projected Lemaître as a victim of those discrepancies that how whole para under equation 23 was replaced by a single sentence where redshift-distance was explained in details. The paper started first from blaming editor of journal and then Hubble or Eddington (teacher of Lemaître) but then concludes that recently it came to surface that Translation was written by Lemaître himself and modifications in translation were his own personal choices.

At that time, I did not read that paper in complete so could not realize that the paper also contains 'right revised' translation in Appendix at the end. I simply rushed to download original French article. Yes there was a complete paragraph under equation No.23 which was replaced by a single sentence in the English

Translation. But since I could not read French article so I typed relevant para in notepad and sought google translation of para. And the resulting translation was depicting a gigantic blunder of Translation of 1931. The revised translation of original French explanation under equation No.23 includes following crucial sentences:

"Radial velocities of 43 extragalactic nebulæ are given by Str omberg (6). The apparent magnitude m of these nebulæ can be found in the work of Hubble. It is possible to deduce their distance from it, because Hubble has shown that extragalactic nebulæ have approximately equal absolute magnitudes (magnitude = − 15. 2 at 10 parsecs, with individual variations ±2), the distance r expressed in parsecs is then given by the formula log r = 0,2m + 4,04."

Actually, 1927 French article was published in an obscure journal and original article had failed to receive attention by scientific community. At that time, Lemaître had sent copy of article to his former teacher Arthur Eddington but he also did not respond and perhaps only had a cursory look of that article. According to para under equation No.23 of the original article, redshift-distance relation was not derived from any relativistic equation but was incorporated in the formulation of equations to get matching results with known observational data of redshift as well as distance. Essentially, redshift-distance relationship was formulated in exact same mode as later on Hubble would also formulate. The redshift-distance relationship had no mathematical derivation − it was simply derived from observational data. And Lemaître was the first to present that relationship but his original work did not reach to the right audience. So far there was no blunder. Only thing was that accompanying relativistic equations themselves never gave result of expansion but without proper experimental basis, Lemaître assigned meanings of 'expansion' to the whole new type of cosmological redshift which he discovered himself.

After two years, Edwin Hubble would find the same redshift-distance relationship in same (non-mathematical) observational data mode but he would not commit mistake of blindly assigning the meaning of expansion despite having no observational proof that redshift-distance actually had anything to do with receding of anything from observer. Rather, he would actively search for right person who could provide satisfactory theoretical justification for galactic redshift-distance relationship. However, everyone will listen to him only up to the statement that there is "(apparent) velocity-distance relationship" and everyone will automatically understand this statement in a modified form of "velocity-distance" relationship (i.e. automatic omission of word 'apparent'). "Velocity-distance" would acquire the status of a confirmed scientific fact on authority of Edwin Hubble and discovery of expanding universe will be attributed to Hubble despite the fact that Hubble himself would remain skeptical to the idea of expansion and it is also possible that Hubble also sometime be using both terms "redshift-distance relationship" and "velocity-distance relationship" interchangeably.

Thus after 1929, "velocity-distance relationship" was known to everyone as newly found fact by Hubble. Eddington, former teacher of Lemaître, at that time was in an effort to account for observed 'velocities' of galaxies within the framework of relativistic equations[10]. After knowing that Eddington was in search of kind of solution that he developed in year 1927, Lemaître again sent him copy of his paper and this time Eddington overwhelmingly acknowledged his article and also reported to de-Sitter, another prominent relevant mathematician of that time. Perhaps Eddington persuaded Lemaître to write English translation to be published in a reputed journal. Eddington even sponsored the translated article by himself writing a supportive article for same publication. Lemaître became a celebrated scientist in 1931 due to publication of translation of the original article. But there were blunders in translation.

If we read only translation, then it is a magically great article because without any reference to Hubble, there is derivation of a hard fact from equations such that hard fact was to be discovered by Hubble only two years after the publication of original French paper. The greatness of article (translated) was also greatly felt such that soon Lemaître would be invited to great conferences where he would propose as ridiculous ideas as 'Primeval Atom' (later became 'Cosmic Egg') for the whole of Universe and all the celebrated audience would accept like under trance.

But nothing was great up to the magical level. Lemaître had found a simple linear relation on the basis of observed data that he had. That relation was not derived from equations but equations were designed to remain consistent with observed data. No scientist would formulate equations without properly taking care of available observational data. If Lemaître had found that relationship purely out of equations then he should have explained this fact in the original French article. But in the original article, he simply writes that distance is found by applying simple deduction on observational data. And since he knew the redshift-distance relationship out of observational data, he was able to propose 'radiation pressure' as cause of expansion. We have already seen that idea of 'radiation pressure' was not derived from equations. He simply empirically knew the linear relationship between redshifts and distance and he only arbitrarily suggested cause of relationship to be the 'radiation pressure'. After the publication of manipulated translation with the help of Eddington in 1931, he 'abandoned' the associated idea of 'radiation pressure' and did not pursue or develop it further.

In the capacity of a translator, it was his first duty to present only the original article in the translated language. But if any modification was indispensable, then he was duty bound to explain reasons for modification along with presenting translation of omitted portions in footnotes at least. The discrepancy in translation was perhaps never surfaced during his life time and even to-date the discrepancy is not widely known. There have been speculations regarding who omitted crucial parts of the article from translation. At first editor of journal; then Eddington and even Hubble is blamed for the omission. Lemaître has been projected as victim of the discrepancy as it deprived him of priority claim in finding Hubble law. Those who understand the meaning of French paragraph take it only from the point of view of who first time discovered expanding universe; Edwin Hubble or Georges Lemaître. The issue is largely overlooked from angle whether the relationship was derived from equations or equations were framed according to available observational data. However speculations regarding who omitted crucial paragraphs from translation have been resolved through special efforts of Mr. Mario Livio[11] who has found a letter written by Lemaître to the editor of journal where Lemaître is telling the editor that "I did not find advisable to reprint the provisional discussion of radial velocities which is clearly of no actual interest." So it was someone's advice to not include 'provisional discussion' of radial velocities which is of 'no actual interest'. That someone should be Arthur Eddington, his former teacher who also happened to be at authoritative position of Royal Astronomical Society. The journal where translation was to be published was also under the administrative control of Royal Astronomical Society. Mario Livio takes above-mentioned words of Lemaître as his humbleness since he is not showing interest in a priority claim regarding discovery of expanding universe. Well, this could be humbleness or innocence of Lemaître that he was only being guided by his former teacher. Lemaître himself could be blank regarding what was the actual goal of Eddington but he should not be as simple as to call that crucial paragraph as having 'no actual interest'. By all means that was a crucial and interesting paragraph. By choosing not to reprint 'provisional discussion', he did not abandon his priority claim. The linear relationship of redshift-distance was still present at the end of article where numerical results were presented in a table. After omission of 'provisional discussion' that was actually reference to Hubble as a source of that relationship, now the end part of article had become like a manifestation of magic that was showing how only the equations had already derived a hard fact two years before the actual discovery of that fact. Editor of journal raised no objection and published the modified translation. Eddington also wrote a sponsoring article in same issue of journal and while having published a sponsoring article in the same issue, the fact of modified translation could not be out of sight of Eddington.

Before moving on, one thing needs to be settled. Mario Livio writes that in 1927, Lemaître first derived Hubble law from equations and then went beyond mere theoretical calculations and attempted to find actual value of Hubble Constant. In the translation, he only omitted paragraph related to determination of value of Hubble Constant whereas linear relationship of 'velocity-distance' already had been derived from equations. My response is that once you have observational data of 'velocities' (redshifts) and you also know method of derivation of distance, then you can easily suggest linear relationship and there is no need of derivation of linear relationship from complex equations. In fact, Edwin Hubble would actually do the same within next two years. Likewise in 1927, Lemaître had the data of redshifts and he also knew the method of finding distance. He already had a rough sketch of linear relation between redshifts and distance and he simply developed mathematics that was consistent with the available sketch. The omitted paragraph was originally written after equation No.23 of the original French paper and this paragraph included calculation of radial velocity of 625 KM/sec/mega-parsec. This figure has come directly from observational data and it is not even consistent with equation No.23 because in equation No.23, as we shall see in coming paragraphs, H was not a constant term.

While yes, apparently there is resemblance between equation No.23 and the Hubble law which is $V = HD$ whereas equation No.23 is $\frac{V}{C} = \frac{R'}{R} r$.

Here $\frac{V}{C}$ is redshift as Lemaître makes it clear under equation No.22 and also in a given table provided after equation No.31. In Hubble law, redshift is V therefore LHS of Hubble law and equation No.23 of Lemaître is same. Furthermore, D and r of RHS of both equations are also same because both stand for 'distance'. $\frac{R'}{R}$ is change in total radius of universe divided by original radius and this change of radius has occurred in time when light emitted from source (galaxy) has reached to observer.

Now the question is whether $\frac{R'}{R}$ and H of both equations also same? Well, it is not clear but if we accept that equation No.23 is exactly equal to Hubble's law then we should find the source of equation No.23 in Lemaître's article whether it is derived from equation No.22 or from which of the earlier equations?

Apparently however this equation No.23 has come out of nowhere. In equation No.23, r (distance) appears for the first time throughout the article and there is no back source of 'distance' in previous equations. The only source of r (i.e. 'distance') is one sentence written just before equation No.23. The sentence is "When the light source is near enough, we have the approximate formula." It means that Hubble type redshift-distance relationship was not really derived from equations but equation No.23 was formulated to remain consistent with later proceedings where actual data of redshifts and distances of various galaxies was going to be discussed.

Up to equation No.22, there is no reference to 'distance' of light emitting source. Since it is 'Doppler's Effect' interpretation going on, distance is not even relevant because the relevant thing is 'speed'. However, within Doppler's interpretation, there is mention of time of emission of light from source (galaxy) and time when light is observed. With this information, we are obliged to give a remote margin that might be equation No.23 with 'r' was derived from equation No.22. But if it is the case then the linear relationship of equation No.23 resembled to Hubble's linear relationship such that not H, actually D was constant in that equation. The title of section 4 of the article (starting just before eq.22) is "Doppler Effect due to variation of the Radius of the Universe." Even title of the article is "A Homogeneous Universe of Constant Mass and Increasing Radius accounting for the Radial Velocity of Extra-galactic Nebulae." In this scheme of interpretation of redshifts, it is radius of whole Universe which is increasing due to which Doppler's Effect is only 'apparent' i.e. extra-galactic nebulae are at fixed distance and they are not receding away from us.

In section 4 of the article, light is emitted from a coordinate $\sigma 1$ and slightly later another ray of light is emitted from same coordinate $\sigma 1$ and reached to observing coordinate $\sigma 2$. Radius of whole Universe is increased during this slight duration but light emitting coordinate has remained the same. Equation No.22 describes a redshift and just after the equation the text states that "it is 'apparent' Doppler's Effect due to the variation of the radius of Universe." Then equation No.23 presents a Hubble type linear relationship with the crucial difference that instead of H it is D which is constant and 'distance' comes into equations for the first time only out of an introducing sentence. Up to equation No.23, D (or r) is constant but afterwards H (or $\frac{R'}{R}$) becomes constant. Therefore there are two distinct tracks within Lemaître's article. Nice words here do not portray the reality as the fact is that two distinct parts of Lemaître's article are inconsistent with one another.

First part of the article is up to Equation No.23 where at the end, suddenly 'r' (distance) arrive in equation. This distance was constant as the only changing entity was radius of whole universe which is the radius of curvature of universe. The coordinate $\sigma 1$ i.e. light emitting point remains the same but radius of universe changes and Doppler's Shift was only apparent as source of light was not moving – only radius of universe was expanding. Now it is crucial to point out that 'FLRW' metric has picked only Equation No.23 from first part of the article and that also in modified form.

The position of Equation No.23 that coordinate of light emitting source does not change is consistent with 'FLRW' metric where coordinates of receding galaxies also do not change as coordinates themselves recede away. If 'FLRW' principle is to be followed then coordinate of radius of universe also should not change. But within Lemaître's equation, coordinates of radius of universe do change. Moreover, Lemaître is only talking about expansion of curvature of whole universe and he is not talking about 'expansion of space' within 'FLRW' type meanings. In fact, equation No.23, if written like Hubble's law would be given as $V = \frac{R'}{R}D$ but 'FLRW' metric would make this equation into $V = \frac{S'}{S}D$ where S stands for 'space'. But Lemaître had derived his own equations and not FLRW metric. Up to equation No.23, galaxies are not even moving away. Coordinates are also not moving within Lemaître's article. Distance is perfectly constant for the light source. Doppler's Effect is only apparent and it is due to expansion of whole universe. In other words, galaxies are not moving away but somehow gravitational hold of the whole universe is becoming weaker due to which spatial curvature of whole universe is getting straighter thus 'apparent' Doppler's Effect is accounted for in this way. If equations had derived anything then it was this something. Top of all, he was able to derive equations merely because he had the data of 'Doppler's Effect' of various extra-galactic 'nebulae' and he also knew how to deduce distance of those nebulae out of a method which he had learned from Hubble. Obviously he did not learn that method after having derived equation No.23. Given the fact that he already had sketchy idea of linear relationship of redshift and distance, why and how could he formulate structure of equations that should be devoid of this relationship? Evidently, his equations had to be consistent with sketchy idea of empirical facts which he had found himself.

After equation No.23, Lemaître joins a different track. Now he would do actual calculations of redshifts and distance purely out of observational data but this part would later on be omitted in the translated article. After presenting the table under equation No.31, he would first time say that not only radius of Universe but r i.e. distance of galaxy and σ_1 i.e. light emitting point (coordinates) are also proportional to Doppler's Effect. This part of the article would therefore be inconsistent not only with first part of the same article, it will also be inconsistent with famous 'FLRW' metric because in second part of the article, galaxies are physically receding away and not within the meaning of 'expansion of space' since coordinates of light emitting sources are also changing.

Thus my response to Mario Livio is that before the omitted paragraph, Lemaître had not actually reached to Hubble's law which means that GR equations alone failed to take him to the destination of Hubble's law. He reached to Hubble's law after equation No.23 only through the route of observational data and he also determined his original value of Hubble's constant within the framework of observational data mode only.

But in year 1931, he deliberately presented modified translation. With translated article, he projected himself able to derive from equations a hard fact which he actually learnt from observational data and from a method of derivation of distance provided by Hubble. Clearly he was being guided by someone who could most probably be his former mentor – Arthur Eddington, who already had served the role of king-maker by authenticating Einstein's General Relativity through his famous (may be notorious) experiment of confirming bending of light ray during solar eclipse in 1919. Role of Eddington makes sense because modification in translation was in his notice and he was the one who actually comprehended the desired consequences of omission of reference to observational data in the translated article. Eddington was fully aware that omission will highlight extraordinary power of equations that would be beneficial for himself and Lemaître both. The strategy worked. Lemaître never claimed priority in finding redshift-distance relationship. He only let people judge the matter in his favor. Equations received a recognized power in the topic. Einstein (old king) apologized to Lemaître (new king) for not previously accepting genuineness of his work. Expansion became real thing. Einstein also accepted that he had been playing around with fudge factors by abandoning 'cosmological constant'[12] and thus granted permission to the expansion to keep going on. Hubble could not openly challenge expansion regime because he had not received any plausible theoretical justification of Cosmological Redshifts. Not only that Lemaître never claimed priority in finding redshift-distance relation, he also preferred to remain silent on the issue of modified translation of original article. The modification was a mega blunder. If it was not a deliberate manipulation then a clarification should have come from Lemaître which never surfaced. Everyone was giving credit to equations for finding a hard fact yet to be discovered by Hubble but Lemaître never explained that he had learned path to discovery of that fact from Hubble himself. Clearly mathematical equations had no extraordinary power. They do have power only up to the extent of what can be logically deduced from given axioms and parameters. If redshift of Doppler's effect has primary relation only with speed and not with distance and if gravity is concerned with attraction and not with expansion then equations based on these two parameters could not give, except in the way of error, the result of a kind of redshift that has direct relation with distance. It was possible only if equations were erroneously solved or at least one of the parameters already had direct relation of redshift with distance. The original

French article duly acknowledged that parameter but the same crucial acknowledgement was unduly omitted from the Translation. Eddington was fully aware that mathematics does not really possess extraordinary magical powers. But he accepted all the benefits of confirming extraordinary powers of mathematics for the cases of Albert Einstein (1919 solar eclipse verification) and then Georges Lemaître (1931). Perhaps he was contented in serving the role of king-maker for these two persons. But at a later stage, he would not be comfortable in again serving the role of king-maker for the case of Chandrasekhar[13] [14] where he would argue that mathematics alone was not able to find realities of physics.

We started our analysis based on our knowledge of that time that Lemaître had employed Doppler's redshift data within the framework of relativistic equations. Now we know that he had not employed Doppler's redshift data rather had employed Cosmological Redshift data within the framework of relativistic equations. Due to this reason he was able to categorically suggest Expanding Universe because he failed to properly distinguish cosmological redshift from Doppler's Redshift. Whereas Friedmann had not employed redshift data in any form whatsoever i.e. Doppler's Effect or Cosmological Redshift. Therefore he did not categorically suggest expanding Universe; rather he suggested expanding or oscillating models depending on chosen value for cosmological constant. Nevertheless, the question arises how after all expanding model (may be in oscillation form) could be derived solely from equations?

I.V. Without using Doppler's Shift Data and without knowing about Cosmological Redshift, Friedmann had already reached to the concept of Expanding Universe. How?

I have explained earlier that GR equations themselves could not provide lead towards expanding model of Universe. GR equations are field equations whose actual function was only to describe path (or curvature) of test particle under the given strength of mass-energy density. But Einstein pioneered the attempt to develop model of whole Universe solely on the basis of field equations by finding solution to equations by specifying certain assumptions and values for certain parameters. One of his main assumptions was that (i.e. assumption is not derived from equations)

Universe has a finite radius. In his famous 1917 paper[15], Einstein has 'assumed' finite radius of universe in following words:

"From what has now been said it will be seen that I have not succeeded in formulating boundary conditions for spatial infinity. Nevertheless, there is still a possible way out without resigning as suggested under (b). For if it were possible to regard the universe as a continuum which is finite (closed) with respect to its spatial dimensions, we should have no need at all of any such boundary conditions. We shall proceed to show that both the general postulate of relativity and the fact of the small stellar velocities are compatible with the hypothesis of a spatially finite universe; though certainly, in order to carry through this idea, we need a generalizing modification of the field equations of gravitation."

"For if it were possible to regard" – that means the suggestion of 'finite' (closed) universe has not come from GR equations. It was like a commonsense judgment that idea of finite universe will fit into the rest of relativistic postulates and other 'facts' that include "small stellar velocities" etc.

Here we are noticing that in 1917, Einstein is trying to develop a model of universe and although he technically discussed (in first pages) the implications of infinite universe but then he "just assumes" finite universe as a proper case to be proceeded upon. He even announces to bring modifications in field equations only to carry through this idea.

"In order to carry through this idea, we need a generalizing modification of the field equations of gravitation."

With this 'modification', he was going to introduce his famous 'Cosmological Constant'. But what was the need to introduce 'Cosmological Constant'? Well, it was needed because, as Einstein himself shows, original GR equations did not support the 'assumption' of finite universe. In paragraph following the equation No.13, he writes following:

> "We should probably have to conclude that the theory of relativity does not admit the hypothesis of a spatially finite universe."

But Einstein had 'intuitively' made up mind to move on with 'hypotheses' of finite universe and he was ready to modify his equations which he did by introducing Cosmological Constant.

With this assumption of finite universe, Einstein actually realized that gravity shall cause matter to condense. The scenario of contracting universe was the natural and commonsense consequence of the assumption of finite radius. Relativity supporters often boast that GR equations themselves initially 'predicted' expanding or contracting universe that tempted Einstein to introduce cosmological constant in year 1917 to confirm to the accepted point of view of that time. But why relativity supporters not boast these things when they get same disinformation right from NASA's website? Following is a quote from NASA website[16]:

> "The Big Bang model was a natural outcome of Einstein's General Relativity as applied to a homogeneous universe. However, in 1917, the idea that the universe was expanding was thought to be absurd. So Einstein invented the cosmological constant as a term in his General Relativity theory that allowed for a static universe."

Actually GR equations themselves had no 'prediction' at all. It happen that intuitively Einstein thought that let universe be finite. But his own equations did not accompany him. Original equations were neither giving him 'static' nor 'expanding' universe – original equations when coupled with intuitive idea of finite universe were giving him 'collapsing' or 'contracting' universe. Therefore either it is plain misunderstanding or utter lie that original GR equations had the 'prediction' of static or expanding universe.

He evaluated the idea of infinite universe but infinite universe had complications with regards to various postulates of relativity. Therefore he preferred the intuitive idea of finite universe and even modified his equations to pursue that otherwise incompatible idea. To carry through this intuitive idea, Einstein introduced a clear fudge factor in equations in the form of cosmological constant. We are told that this cosmological constant physically represents energy density of the vacuum of space[17] which exerts anti-gravity type repulsive force which does not let universe to contract. Here, my objection on physical meaning of cosmological constant is that energy-density (even if it is of 'vacuum of space') should add to more gravity rather than giving any sort of anti-gravity. But anyhow, accepted meaning of cosmological constant was anti-gravity whose parametric value could cause expansion, static stability or contraction. In short, possibility of 'expansion' was provoked solely out of a commonsense assumption of a finite radius of universe such that the assumption was not derived from equations.

In 1922, Friedmann showed that zero parametric value of cosmological constant will give the result of a stable oscillating universe with oscillating period of 10 billion years if mass contained in the universe is 5×10^{21} solar masses. Therefore by using cosmological constant in his equations, Friedmann made a commonsense assumption as part of his mathematical analysis. Furthermore, he added his own assumptions also. The Universe of Einstein had definite radius and the length of radius was dependent on quantity of (finite) matter contained in the Universe. Perhaps at the time general estimate of total mass content of Universe was the same figure of 5×10^{21} solar masses. The radius of Einstein's static Universe had no relation with time as radius had relation only with total mass content of Universe. Here Friedmann added another assumption from outside the realm of equations. He added the 'assumption' that radius of Universe was dependent on time (radius was function of time). In the translation of his famous 1922 paper, Friedmann describes mathematical model of Einstein's Universe in following words:

"Einstein obtains the so called cylindrical world, in which space possesses a constant curvature independent of time and in which the radius of curvature is connected with the total mass of matter existing in space."

It is clear from above quote that relationship of radius with time could not be derived from mathematical model as proposed by Einstein. Afterwards, Friedmann tells us the goal of his own work which includes following:

"Second (goal is), the proof of the possibility of a world whose spatial curvature is constant with respect to three coordinates that are permissible spatial coordinates and that depend on time, e.g. on the fourth (time) coordinate. This new type is, as far as its remaining properties are concerned, an analogue of the Einsteinian cylindrical universe."

Here we see that actually Friedmann is going to develop a new type (of model) which would be outside the framework provided by Einstein's model. Within the framework provided by Einstein's model, spatial curvature does not depend on time. But Friedmann wants to prove possibility of a separate kind of world where spatial curvature would depend on time. This is very important point to consider because modern Big Bang Cosmologists always tell us about the supremacy of GR equations that they already secretly contained, without being in notice of Einstein, the super powerful concept of 'singularity' from which our whole universe has been originated. But when radius of Universe had no relation with time under GR equations then backward in time projection of radius of Universe at time 0 as 'singularity' was also simply nowhere in GR equations.

Anyhow, Friedmann proceeds to describe two classes of assumptions for his own model. The first class of assumptions coincided with the assumptions of Einstein and de-Sitter (de-Sitter also had developed solution of GR equations for a model of Universe). The second class of assumptions was new comer and had no relation with previously developed models. The crucial assumption under second class as narrated by Friedmann was "R (radius) depends only on x_4 (time coordinate) and it is proportional to the radius of curvature of space, which may therefore change with time". Here important thing to be noticed is that though Friedmann assumed radius of curvature proportional to time but he has totally skipped first class of assumption according to which radius of curvature should also be proportional to total mass content of Universe. But since he has already mentioned first class of assumptions hence we should conclude that first class of assumptions shall remain valid part of further proceedings. This aspect gets clear under equation No.5 where Friedmann makes it clear that "R (radius of Universe) is a function of x_4 (time coordinate) and M (Total mass content of universe) depends, in the general case (i.e. Friedmann is calling his own model as general case), on all four world coordinates (i.e. three spatial and one time coordinate)".

Now we have reached to very important point. Above analysis is actually making it clear that Friedmann's model of expanding universe is consistent with the Steady State Model of Universe but categorically does not support Big Bang Cosmology. If, with the passage of time, Universe is expanding then total mass content of Universe is also increasing. This position of Friedmann is not in harmony with the Big Bang Model. However, this position is in line with the Steady State Model. On the contrary, Wikipedia article on Alexander Friedmann states that the dynamic cosmological model of 'general relativity' developed by him became standard for both the Big Bang and the Steady State theories. According to this Wikipedia article[18], Friedmann's work equally supported both theories and that Steady State theory was abandoned only after detection of CMBR.

Here first of all I should register my objection on the notion that Friedmann's cosmological model belonged to general relativity (GR). I have explained it earlier that cosmological constant was not derived from GR equations but was simply assumed as a commonsense based consequence of non-mathematical assumption that Universe has finite radius. GR equations themselves could not give result of either expansion or contraction. Only with an extra assumption of 'finite radius of universe', the need for a fudged solution evoked. Expansion was mathematical consequence of this type of fudge factor. This fudge factor cannot be fully traced back to GR equations. This fudge factor can be traced only up to a commonsense assumption and resultant commonsense solution. To register this objection was crucial because such instances highlight how relativists unduly trace every aspect of the Big Bang theory to GR equations and try to demonstrate superiority of (mathematical) equations in general and GR equations in particular. The only proof of Big Bang is actually like this – Hubble found expanding Universe >>> Expansion was already derived from GR equations by Friedmann (1922) and Lemaître (1927). We people now keep on saying that what Hubble had found was not 'expansion' as Cosmological Redshift which he had found is not the proof of expansion of space or physical receding of anything. While we may keep on saying this, relativists keep on saying that since Hubble's findings already had been derived from (GR) equations, therefore there is no need to physically demonstrate that meaning of Cosmological Redshift is anything other than 'expansion of space' as depicted in Friedmann-Lemaître type equations.

After having registered the above objection, now we come back to the main discussion. We have seen so far that Friedmann's model is actually not consistent with the Big Bang Theory however there is supportive material for the Steady State Theory. The Big Bang Theory and the Steady State Theory are the only two accepted theories under standard model because both theories accept and adhere to basic framework of 'expanding universe'. Both theories accept that Cosmological Redshift, even before having been discovered, was already mathematically described in terms of Expanding Universe by Friedmann and Lemaître. Wikipedia article[19] defines Steady State Theory in following words:

> "In cosmology, the Steady State theory is an alternative to the Big Bang model of the evolution of our universe. In the steady-state theory, the density of matter in the expanding universe remains unchanged due to a continuous creation of matter, thus adhering to the perfect cosmological principle, a principle that asserts that the observable universe is basically the same at any time as well as at any place."

Thus Friedmann's model is actually supporting the Steady State theory because (i) radius of universe expands with time and; (ii) Total mass content of Universe also increases with increase in radius and in this way total mass density of the Universe remains the same. But standard model has the claim that Friedmann's model actually supports both (i) the Steady State and; (ii) the Big Bang theories equally. But – the standard Big Bang theory does not permit continuous creation of more mass with the ongoing expansion.

Now we come back to original 1922 paper of Friedmann where he starts part-II (B) of the paper with sentence, "We now want to consider the non-stationary world. M (total mass content of universe) is now function of x_4 (time coordinate)". We see here that for Friedmann, dynamic universe is not just contracting or expanding in terms of radius, it is also losing or gaining mass. But more relevant to Big Bang points are yet to come in Friedmann's 1922 paper. For the derivation of equation No.20, he writes:

"Since the radius of curvature may not be smaller than zero, it must decrease with decreasing time, t, from R0 to the value zero at time t'. We shall call the growth time of R from 0 to R0 the time since the creation of the world".

With above in the celebrated 1922 paper of Friedmann, we have actually reached to the basic idea of Big Bang. Friedmann calls the world at time zero as 'monotonic world of the first time'. Story does not end here. Under footnote No.11, Friedmann writes following:

"11. The time Since the creation of the Universe is the time that has elapsed from the moment when space was a point (R=0) to the present state (R= R0): this term may also be infinite."

What we have found here are the plain original ideas of 'singularity' as well as 'expansion of space'. Now we shall analyze these two aspects right here. First of all let us emphasize here that Lemaître also had learned his basic idea of 'Primeval Atom' or 'Cosmic Egg' from these points which are contained in famous 1922 paper of Friedmann. A pro-Friedmann paper confirms this point in following words:

> "In 1931, Lemaître first gave Friedmann's singularity a physical meaning, that of a "primeval atom" blowing up—what Fred Hoyle later dismissively called "the Big Bang."20

Story emerged so far is that while in year 1927 Lemaître had proposed expansion of universe but by that time he was unaware of Friedmann's work. Einstein, while rejecting 1927 work of Lemaître, had told him that similar expanding universe solution was already presented by Friedmann. After publication of manipulated translation of 1927 article in 1931 by Lemaître, Einstein publically abandoned his concept of cosmological constant. In fact he had not abandoned this concept altogether but had conceded to the value assigned to it by Friedmann. These developments compelled Lemaître to review his own work in the light of Friedmann's ideas. Thus, in 1931, he picked the idea of 'monotonic world' from Friedmann, assigned physical meanings to it and called it 'Primeval Atom' or 'Cosmic Egg'.

It is now clear that concept of 'initial singularity' has come in the standard Big Bang Cosmology from Friedmann. First thing is that since Friedmann's expansion or oscillating model was based on 'cosmological constant' which was not the part of original GR equations, therefore the idea of initial singularity also has nothing to do with GR equations. Second thing is that concept of 'initial singularity' has come from an incorrect understanding of actual model of Friedmann. The actual model of Friedmann is based on two classes of assumptions. First class of assumptions includes the assumption that radius of universe is function of total mass content of universe. Second class of assumptions includes the assumption that radius of universe is a function of time. While specifying second class of assumptions, Friedmann has used word 'only' with the word x_4 (i.e. time). The usage of word 'only' has deceived Big Bang Cosmologists into believing that radius of universe is function of time only. Here they completely forget that Fiedmann also has specified another class of assumptions where he has assumed that radius of universe is function of total mass contents of universe as well. Now the standard concept of 'initial singularity' of standard Big Bang Model is based on exactly this mistake. This view is further strengthens on account of the fact that in the translation note of the translation of 1922 paper, the translator also has committed the same mistake. In the translation note[21], the translator has written following:

> "If R (radius of universe) is independent of time, then the stationary world models of Einstein and Wilhelm de-Sitter follow. If R(t) depends only on the time variable, then a variety of monotonically expanding or periodically oscillating models result, depending on the value chosen for cosmological constant."

What we need to understand here is that Friedmann has not actually presented any expanding model of universe. What he has presented is a broader and general scheme of all the mathematical possibilities of stationary as well as non-stationary models. Expanding or oscillating models are only particular cases of this general scheme. Even stationary models are also particular cases of this general scheme. More precisely, two particular stationary models of Einstein and de-Sitter[22] were available by his time. Stationary model of Einstein was cylindrical universe model where radius depended on mass content only. Wilhelm de Sitter's spherical universe model was more geometrical where even mass content was also not discussed. After identifying the nature and type of available stationary models, then Friedmann proceeds to formulate a general scheme. The general scheme shall cover both stationary as well as non-stationary models. But the whole general scheme would be based on two classes of assumptions. In the new general scheme, stationary models will follow both classes of assumptions and non-stationary models also would follow both sets of assumptions. In this way, Friedmann, at first, was going to amend already available two stationary models. In equations No.6 to 10, Friedmann thus derived Einstein's model and de-Sitter's model separately such that now these two models were based on both classes of assumptions and in this way Friedmann made it clear that both the stationary models of Einstein and de-Sitter were basically special cases of his own general scheme which was based on two classes of assumptions. After equation No.10, Friedmann proceeds to consider the non-stationary worlds and clearly writes "M (total mass content of universe) is now a function of x_4 (time coordinate)". With this sentence, every doubt should be cleared. His non-stationary models were dependent on both classes of assumptions and not on time coordinate only (i.e. only second class of assumptions).

I.VI. Concept of 'Initial Singularity' of Modern Big Bang Cosmology has been derived from Incorrect

Understanding of Friedmann's Model

Friedmann has presented a general scheme of stationary as well as non-stationary models of universe. Both types of models are based on two classes of assumptions. First class of assumptions included that radius of universe is function of total mass contents of universe. Second class of assumptions included that radius of universe is function of time. Then Friedmann started discussing possibility of 'monotonic world' i.e. world at time zero and radius zero. Here Big Bang Cosmologists committed a crucial mistake and made whole universe into a mythological fiction of zero radius with infinite density of mass. They simply ignored first class of assumption that radius was dependent on total mass content as well. If Friedmann is discussing possibility of a monotonic world where radius of universe is zero at time zero, then total mass content of universe was also zero at that zero time. Its meaning is that in mathematics, there is no valid concept of infinitely dense mass within infinitely small point and thus 'initial singularity' concept of modern Big Bang Cosmology is nothing more than an incorrect fiction. Friedmann was presenting only abstract mathematics where he ruled out possibility of certain scenarios only due to one reason that square root under that option was imaginary number. Zero was not a reason, within mathematical analysis, to rule out possibility of a scenario. Zero space with zero mass was thus a valid option within abstract mathematics. But Big Bang cosmologists mistakenly took it as zero space with infinite mass or density, assigned it literal physical meanings and started calling it initial singularity that started to 'expand' with the start of time. Those Big Bang Cosmologists failed to see error messages notified to them by their own commonsense because they were devotees of 'counter-intuitive' physics which was based on ultra-superior 'relativistic' equations. Now, within the right meanings of 'monotonic world' of Friedmann, these (Big Bang) cosmologists are under obligation to tell us about the valid physical processes that can keep on producing new mass after passage of time from initial zero values of both mass and time.

I.VII. Expansion of Space

The case of the Big Bang Cosmology is that after the discovery of Hubble's law in 1929 that 'more distant galaxies are moving away at greater speed', scientists realized that this law was already derived from GR equations by Friedmann (1922) and Lemaître (1927). We have seen already that Lemaître had actually found this law in 1927 out of observational data and he did not derive it from any equation. For the case of Friedmann, let us now analyze whether he actually derived this law in 1922 or not. But before analyzing this aspect, let us first confirm the case of Big Bang Cosmology as mentioned above. The following is written in Wikipedia article on Hubble's Law[23]:

> "Although widely attributed to Edwin Hubble, the law was first derived from the general relativity equations, in 1922, by Alexander Friedmann who published a set of equations, now known as the Friedmann equations, showing that the universe might expand, and presenting the expansion speed if this was the case. Then Georges Lemaître, in a 1927 article, proposed the expansion of the universe and suggested an estimated value of the rate of expansion, which when corrected by Hubble became known as the Hubble constant."

Before attempting to find this law in Friedmann's 1922 paper, it is necessary to understand that modern concept of 'expansion of space' is deeply linked with Hubble's law. Although I have objection on usage of term 'velocities' in Hubble's law as Hubble has only noted relation of 'redshift' with distance and not 'velocity' with distance and he had clarified that he had used term '(apparent) velocities' but let us move on with the term 'velocities' because the same is the accepted meaning under standard model. So within the standard meaning of Hubble's law, the first problem aroused then 'why do we appear to be at center?'

This problem was resolved easily by using expanding balloon surface analogy as every point on balloon surface would experience that every other point is moving away from it and every point could take itself at center. The second problem was that Cosmological Redshift (redshift-distance relationship) was not the physical proof of receding of anything. Third problem was that if more distant galaxies are receding away with greater speed then the galaxies located at far off astronomical distances must be receding away at speed greater than speed of light which is not permissible under the same standard model. The 'solution' for the second and third problem was this idea of 'expansion of space'. Cosmological Redshift is not the physical proof of receding of anything but idea of expanding universe is rescued through this idea of 'expansion of space'. Galaxies are not physically moving away from us. It is actually 'space' which is expanding everywhere at constant rate which corresponds with Hubble's constant. And the proof of 'expansion of space' is Friedmann- Lemaître equations. Since galaxies are not physically moving away as only space is expanding so there is also no actual problem of receding speed greater than speed of light.

Now we come to the 1922 paper of Friedmann to see extent to which it is true that Hubble's law was already derived by him through equations or was he really talking about 'expansion of space' within the modern standard meanings of this notion.

The expanding universe model of Friedmann is that radius of universe expands with passage of time and creation of new mass. Zero radius at zero time may reach to maximum radius in 10 billion years with total mass of 5×10^{21} solar masses. If more mass is not created then total mass will start diminishing and in next 10 billion years, the radius and mass quantum both will again reach to zero. Now readers are invited to judge by themselves regarding where is Hubble's law in this type of expansion model? In this expansion model, continuous induction of new mass is required. It is not Hubble's Law of experimental physics. This is Friedmann's law of Abstract Mathematical Physics. Now suppose that time is passing and mass is being created at uniform rate, then speed of expansion of radius will also be uniform. When radius is 1, expansion speed is 100. When radius is 13 billion light years, expansion speed is again 100. This is not speed-distance relationship of Hubble's law. It is not even speed-mass relationship. Hubble type expansion is possible only if every second, greater than the previously added mass is created. If at first second 1 Kg mass is created and the same increment of 1 Kg is being created every next second, then it is not the case of Hubble type expansion. But if at first second 1 Kg mass was created, at second increment was 1.1 Kg and at third second increment was 1.2 Kg, then it would be a proper case of Hubble type expansion. But Friedmann had not made equations for these things. When he has calculated time period of 10 billion years for mass of 5×10^{21} solar masses he has not even told the value of radius after 10 billion years or that what was mass and radius after let's say 5 billion years. In no way could Friedmann found Hubble's law in 1922 on the basis of mathematics alone and neither did he found. Claim of Big Bang Cosmologists that he already had derived Hubble's law from GR equations is not hereby accepted. His equations only could give similar to Hubble's Law type graphs but only depending on increasing incremental values of newly created mass with passage of time. And continuous increase in total mass is not a valid or even remote part of standard Big Bang Cosmology. This thing might be relevant to the Steady State Cosmology but Steady State is already defeated theory and therefore is not on the hit list of this book. When we consider the actual fact that Hubble's law does not even talk about speed, then along with Big Bang, Steady State also becomes irrelevant. In addition, if Friedmann really had reached to Hubble law type expansion then he should not have described oscillation model in simple terms. He should have told us that with maximum radius achieved, contraction would be more difficult because expansion had to be at higher speed at maximum radius. In short, in simple terms of Hubble's Law, greater radius means greater recessional velocity then how contraction phase could initiate at all and why Friedmann has described possibility of oscillation model without first removing this difficulty? Fact is only that in 1922, he had not reached to Hubble Type expansion model neither he could reach to this concept solely on the basis of mathematical analysis of GR equations.

Now we move to the issue of 'expansion of space' and find it true that plain (but shallow) reading of Friedmann's 1922 paper does suggest as if he was talking about 'expansion of space'. Following two portions of his 1922 paper, particularly the second one are capable to give idea of 'expansion of space':

"From that, it follows that R is an increasing function of t. The positive initial value R0 is free of any restriction. Since the radius of curvature may not be smaller than zero, it must decrease with decreasing time, t, from R0 to the value zero at time t'. We shall call the growth time of R from 0 to R0 the time since the creation of the world".

"11. The time Since the creation of the Universe is the time that has elapsed from the moment when space was a point (R=0) to the present state (R= R0): this term may also be infinite."

Both these portions if read in isolation can mislead us into thinking that radius of universe is function of only time and not the function of mass contents of universe. But we have already seen that both first and second classes of assumptions are valid feature of the overall general scheme of possible models presented by Friedmann. Therefore R is function of t and R is also function of M. But here Friedmann is discussing only two variables R and t. A valid assumption 'M' is not being assumed at all. When a valid thing 'mass' is not even being considered then we have to accept that yes he is actually talking about 'expansion of space'. We must consider another aspect also that Friedmann is discussing things within the framework of Abstract Mathematics only. Mathematics is study of space (dimensions, area, volume, shape etc.) and numbers (real, unreal, constants, variables etc.). Within a mathematical model, Friedmann is discussing about space. We must not conclude that he has made 'space' into a real thing having a solid object like capabilities of expansion or contraction.

At this point, we must try to understand Friedmann's actual concept of space. The English Translated title of his 1922 paper is "On the Curvature of Space". By the term 'radius of universe' his meaning is that mass contents of universe would cause gravitational boundary of universe that a straight line universal journey of a physical object would be a complete circle and would reach back to the original point. 'Radius of universe' is radius of this universal 'straight' line which is actually circular.

Within this meaning of 'space', it is physically valid to say that space may expand or contract. Within mathematical model of Friedmann, space is really expanding or contracting according to this meaning. Following are some examples in Friedmann's paper of usage of term Radius R as curvature of space:

"Here R depends only on x4 and it is proportional to the radius of curvature of space, which may therefore change with time."
While deriving constant universe model of Einstein within his own general scheme, Friedmann writes: "whereby R signifies the constant (independent of x4) radius of curvature of space."
"If we restrict our consideration to positive radii of curvature".
"Let the radius of curvature equal R0 for t = t0."
"Positive or negative depending on whether the radius of curvature is increasing or decreasing for t = t0."
"by choice of the time it can always be arranged such that the radius of curvature increases with increasing time at t = t0."

It is now clear that yes space is contracting or expanding in Friedmann's model but it is contracting or expanding within above physically valid meanings of contraction or expansion of space. But Big Bang Cosmologists tell us a whole different and misleading thing and they attribute their own faulty model to Friedmann. They call their own misleading model of 'expansion of space' as 'metric expansion of space' and wrongfully attribute this faulty physical model to Friedmann. Following are the accepted meanings of metric expansion of space according to Wikipedia article[24]:

"The metric expansion of space is the increase of the distance between two distant parts of the universe with time. It is an intrinsic expansion whereby the scale of space itself changes. It means that the early universe did not expand "into" anything and does not require space to exist "outside" the universe - instead space itself changed, carrying the early universe with it as it grew. This is a completely different kind of expansion than the expansions and explosions seen in daily life. It also seems to be a property of the entire universe as a whole rather than a phenomenon that applies just to one part of the universe or can be observed from "outside" it. Metric expansion is a key feature of Big Bang cosmology, is modeled mathematically with the Friedmann-Lemaître-Robertson-Walker metric and is a generic property of the universe we inhabit. However, the model is valid only on large scales (roughly the scale of galaxy clusters and above), because gravitational attraction binds matter together strongly enough that metric expansion cannot be observed at this time, on a smaller scale."

So the article is proudly saying that this model is valid (or physically detectable) only on large scale astronomical distances. Whereas as per Friedmann's actual model if universe consists of only 1 solar mass, then it will have a radius of curvature which will be set by the gravitational boundary of only one solar mass and in physical terms it may be equal to only few thousand astronomical units. In simple terms, it should be equal to largest possible orbit around sun. If universe contains 5×10^{21} solar masses, then radius is beyond of our reach. But standard model is saying that only after local galaxy cluster they are able to see expansion of this radius. Off course they are not able to see expansion of radius as the only thing which they see is 'receding' of galaxies. But Friedmann is talking about increase in radius due to increase in mass and he is not talking about physical receding of galaxies in terms of misinterpreted Hubble's Law. FLRW metric where 'F' stands for 'Friedmann' is only a deliberate modification or at worst, the plain misunderstanding of Friedmann's actual model. Only thing is that science community learned an amazing thing in 1929 that there is a linear relationship between distance and redshift of light coming from far off galaxies. They misread the actual fact in the modified form that there is linear relationship between distance and receding velocities of galaxies. They also wrongfully realized that in year 1922, Friedmann had derived exact this fact from equations of General Relativity. Then two new mathematicians 'R' (Robertson) and 'W' (Walker) might have modified equations of 'F' (Friedmann) and 'L' (Lemaître) and the resultant new metric equations are now known as FLRW metric. This FLRW metric is considered, under standard model, as the only possible explanation of Cosmological Redshifts discovered by Hubble in 1929. There is no physical proof that cosmological redshift has anything to do with physical receding of anything. It is only account of authority of (dubious) mathematics (FLRW metric) that Big Bang Cosmologists do not feel the need to have physical proof that cosmological redshift really means receding of galaxies from us. They do not need any proof and they do not offer any proof. Yet they say that Big Bang is a scientific theory and they promote this clearly false theory as such. Science has been wrongfully disconnected from real observations or experiments and is now based on mathematics. Mathematicians now float their equations in market (official papers) and wait for the time when any real observation would be found remotely consistent with their equations. Then they would jump in with claims that such and such observed fact was already 'predicted' by their equations and sadly, this is the only permissible way of proposal and acceptance of new scientific ideas under the established system of scientific methodology.

Anyhow, we have seen that Friedmann has only presented abstract mathematics. The physics behind expanding model of Friedmann is set out by 'cosmological constant' which is not the genuine part of General Relativity equations. Einstein himself writes following in his 1917 paper where he presented his stationary model of universe by introducing 'cosmological constant':

"In order to arrive at this consistent view, we admittedly had to introduce an extension of the field equations of gravitation which is not justified by our actual knowledge of gravitation. It is to be emphasized, however, that a positive curvature of space is given by our results, even if supplementary term is not introduced. That term is necessary only for the purpose of making possible a quassi-static distribution of matter, as required by the fact of small velocities of the stars."

Second thing is that Firedmann did present expanding model but a variable curvature of space depending on time and mass was not out of sight of Einstein in 1917:

"Curvature of space is variable in time and place, according to the distribution of matter, but we may roughly approximate to it by means of a spherical space."

However here Einstein might be talking about curvature of space at particular location of universe. Friedmann extended this idea to the curvature of whole universe. But neither Einstein (up to that time), nor Friedmann (ever) talked about 'FLRW' metric type expansion of space which is causing far off galaxies to move away from solar system at speeds greater than speed of light. In fact, one of the fundamental assumptions of Einstein, in year 1917, was that speeds of stars are too low as compared with velocity of light. In 1917 paper, he wrote following:

"We shall proceed to show that both the general postulate of relativity and the fact of the small stellar velocities are compatible with the hypothesis of a spatially finite universe."

"The most important fact that we draw from experience as to the distribution of matter is that the relative velocities of the stars are very small as compared with the velocity of light."

Thus we see that, while not knowing Hubble type expansion in year 1917, Einstein could think of local variable curvature of space that depended on time and distribution of matter. In 1922, Friedmann was also equally unaware of Hubble type expansion and he could think of variable curvature of space for the whole universe. Friedmann never challenged the 'fact drawn from experience' that relative velocities of stars are very small as compared with the velocity of light. If he (Friedmann) knew anything about coming 'FLRW' metric then he should have explained in 1922 that though relative velocities of stars are very small as compared with the velocity of light but 'proper distance' between heavenly objects is increasing at speed greater than the speed of light due to 'FLRW type expansion of space'. But actually he did not explain this crucial difference of his model with Einstein's model. He only stated that Einstein's model was a special case of his own general scheme. To derive case of Einstein's model within the framework of his general scheme, he never stated that heavenly bodies must move apart at enormous speeds. Within his general framework, he reached to the same model of Einstein with no modification of idea of Einstein that stars have very low relative velocities. In fact, if Friedmann had really reached to the fact of Hubble Type expansion, then his whatever 'general scheme' should not have accomodated the stationary models of Einstein and de-Sitter as special cases. Fact is only that 'FLRW' metric is not consistent with the actual Friedmann and 'FLRW' metric is only an after development when Hubble's Law had already been surfaced.

Now what we see in 1922 paper of Friedmann is that he also has assumed very low relative velocities of heavenly bodies. Under serial No.2 of the first class of assumptions, he writes following:

"The matter is incoherent and relatively at rest. Stated less strongly, the relative velocities of matter are vanishingly small in comparison with the velocity of light."

We know that first class of assumptions, just like second class of assumptions, form the core framework within which whole general scheme of possible stationary as well as non-stationary models of universe operate. If, for Friedmann, relative velocities of heavenly bodies are vanishingly small in comparison with the velocity of light, then 'expansion of space' for him is only expansion of overall curvature of space due to increase in quantity of total matter of universe. If all the matter is relatively at rest, then there is no 'FLRW' type expansion of space going on which is causing matter to relatively move apart at enormous speed that eventually, due to enormous increase in relative distance, crosses the light speed limit. It also means that Hubble type 'expansion of universe' was nowhere in the mind of Friedmann as he did not write another third class of assumptions where he could accommodate enormous relative velocities of heavenly bodies due to 'Hubble' or 'FLRW' type 'expansion of space'.

I.VIII. Differences of Friedmann and Lemaître

Big Bang Cosmologists also try to authenticate their model on account of the 'fact' that after death of Friedmann (1925), Lemaître had independently derived same solution to GR equations as Friedmann presented in year 1922. This 'fact' gives a solid feel about accuracy of mathematics but contrary to Friedmann's clear assumption of vanishingly small relative velocities of heavenly bodies, the same is not the assumption of Lemaître as one of the underlying pillars of his 1927 article was Doppler's shift or velocities data of extra-galactic nebulae. Not only had he employed the available data of recessional velocities of extra-galactic nebulae, he also knew how to deduce distance of those extra-galactic nebulae from a method that he had learnt from Hubble and duly acknowledged this fact in his original French article of 1927 but perhaps on advice of Eddington in 1931, omitted this crucial fact in his translated article. In this way, he was roughly aware of 'velocity-distance' relationship of extra-galactic nebulae. But however, even after knowing the relationship, he has not actually discussed velocities of very far off galaxies whose velocities might cross the speed of light. Secondly, task of Friedmann was to present a general scheme of possible models of universe such that both stationary as well as non-stationary models of universe were possible but Lemaître has not presented a scheme of different possible options as he has presented only the expanding model. The reason for the better clarity is the fact that he was roughly aware of velocity-distance relationship of extra-galactic nebulae while he had not learned or derived this fact from GR equations. There may be a sort of similarity with Friedmann because Lemaître is also talking about radius of whole universe (as curvature of space). But we see that Lemaître is the first person who links redshift data, or better to say, 'velocities-distance' data with the idea of expansion of universe. The expanding model of Friedmann states that with the homogeneous universe but variable time and mass, the radius of universe may contract or expand. But the expanding model of Lemaître is that with homogeneous universe and constant mass, universe expands due to radial velocity of extra galactic nebulae. The English translation of his 1927 article is "A Homogeneous Universe of Constant Mass and Increasing Radius accounting for the Radial Velocity of Extra-galactic Nebulae". Thus the crucial difference of Friedmann and Lemaître is that mass is variable for the former but constant for the later. Friedmann's model has nothing to do with standard Big Bang Model whereas Lemaître's model is the actual start of present day standard Big Bang Cosmology which later on also incorporated misunderstood elements from Friedmann's model.

I.IX. Why after 1929, Scientific Community Misread 'Redshift-Distance Relationship' found by Hubble as 'Velocity-Distance Relationship'?

Well, we have stated earlier that 'velocity-distance' relation was first derived by G. Lemaître from experimental data whose 'velocities' component' had come from Str omberg (may be via Vesto Slipher) and method of derivation of distance was taken from Edwin Hubbel. Lemaître presented this relationship in only one paragraph in his 1927 French article without properly presenting available data. He knew this fact prior to 1929 but he did not well present this fact. His source of knowledge was also the same Edwin Hubble who himself was eventually going to better present this fact by properly making it a case before audience, supplying them data and deducing results therefrom.[25]

There is also no element of 'misreading' by anyone because title of Hubble's 1929 paper was "A relation between distance and radial velocity among extra galactic nebulae".

But even then credit of finding this relationship goes to Hubble. Not only this, science community also misread his actual finding.

The story developed in a way that scientists like Vesto Slipher[26] had been noticing redshifts in spiral galaxies (then considered nebulae) since 1912. Except for few galaxies relating to local group, all galaxies studied by then were found redshifted. Naturally those redshifts were being interpreted in terms of Doppler's Effect and were also being described in terms of 'radial velocities of extra galactic nebulae'. Hubble also employed the same terminology in his 1929 paper. Here in the title of his paper, Hubble has used common term 'radial velocity' for 'redshift'.

It is also true that he starts his paper with sentence "Determinations of the motion of the sun with respect to the extra-galactic nebulae".

But – in the very first paragraph, he is pointing something perplexing for which he is using word 'paradox' and after pointing out this 'paradox', now he is using term 'apparent' velocities instead of velocities.

Following is relevant sentence in the first paragraph:

"Explanations of this paradox have been sought in a correlation between apparent radial velocities and distances, but so far the results have not been convincing."

This is the point. In Doppler's Shift, there is no redshift-distance relationship but what Hubble was observing was a redshift-distance relationship. Therefore with this paper, he was seeking explanations for this 'paradox'.

From whom he was seeking explanation?

He wrote this question in a public paper so he asked it from wise community. But he was also concerned to obtain this explanation from particular people of his choice who were in a position to give authoritative opinion. Furthermore, he has skeptically concluded this paper with following words:

"In the de Sitter cosmology, displacements of the spectra arise from two sources, an apparent slowing down of atomic vibrations and a general tendency of material particles to scatter. The latter involves an acceleration and hence introduces the element of time. The relative importance of these two effects should determine the form of the relation between distances and observed velocities; and in this connection it may be emphasized that the linear relation found in the present discussion is a first approximation representing a restricted range in distance."

In the letter to de-Sitter, he writes:

> "Mr. Humason and I are both deeply sensible of your gracious appreciation of the papers on velocities and distances of nebulae. We use the term 'apparent' velocities to emphasize the empirical features of the correlation. The interpretation, we feel, should be left to you and the very few others who are competent to discuss the matter with authority.27"

Thus 'apparently' redshifts seemed like velocities but for Hubble, the actual interpretation of redshifts was unresolved question. With this, he actually rightly recognized the fact that redshift involved in extra galactic nebulae was of a different kind than to the usual Doppler's Shift. Due to having recognized this fact, the credit of finding redshift-distance relationship rightly goes to Hubble. Now mere fact that Lemaître had found same relation also means that he had reached to the same truth in year 1927 is not true because he had not realized that it was due to a different kind of redshift than usual Doppler's Effect. At a stage, he was also taking 'Doppler's Effect' as 'apparent' in the sense that physical movement was not involved; only radius of universe was expanding. But in the later part of his article, he proceeded with the physical meanings of Doppler's Effect where galaxies had physical receding velocities. Hubble presented the apparent meanings of receding but with due acknowledgement that it was only apparent meaning and the real meaning or explanation had yet to come. In this way, scientific community did not receive or absorb the actual message and took apparent for real. Then in 1931, with the help of Eddington, Lemaître published translation of his earlier article where he omitted crucial parts of original article such that translation was showing as if he already had derived Hubble Type redshift-distance relationship solely from GR equations and without using any observational data. In the translation, the para under equation No.23 was replaced by a single sentence. The original French para duly acknowledged that velocities of extra-galactic nebulae data were taken from Str omberg whereas method of finding distances was taken from Hubble. The translated article was devoid of this crucial information yet in the last column of the table given in the end, the calculated redshifts of extra-galactic nebulae were directly proportional to their distances. It was like demonstration of magic, better to say – show of trick, that how GR based mathematics alone had been able to derive those results in year 1927 which Hubble, on the basis of observational data was able to present in year 1929. Then Robertson (R) and Walker (W), another two mathematicians, entered the scene to modify and/or at-least authenticate Friedmann (F) and Lemaître (L) in the light of recently found new type of redshift whose different features from Doppler's Effect had been noticed and in this way FLRW metric was declared as the only possible explanation of cosmological redshift.

Hubble had carefully used the term 'apparent' velocities to signify that redshift could be due to anything other than 'velocity' at all. He even mentions apparent slowing down of atomic vibrations or a general tendency of material particles to scatter. But 'FLRW' metric provided him the gift of 'literal' or 'real' meaning of 'apparent' velocities. The well-known literal meanings of 'apparent' velocities of far off galaxies are beautifully described in following words in a published paper[28] of Indiana University:

"Two galaxies permanently located at positions (x_1, y_1, z_1) and (x_2, y_2, z_2) at one time find themselves one billion light years apart. Then a few billion years later while located at the same coordinates, they find themselves 3 billion light years apart. The galaxies have not 'moved', nevertheless, their separations have increased. In fact, when the universe was only one year old, the separations between these galaxies were increasing at 300 times the speed of light! Space can expand faster than the speed of light in general relativity because space does not represent matter or energy. The displacements that arise from its dilation produce an entirely new kind of motion for which even our special relativistically-trained intuitions remain profoundly silent. Like that gentleman from Main once said "You can't get there [to general relativity] from here [special relativity]". To the extent that general relativity has been tested and found correct, we have no choice but to accept its consequences at face value."

So Hubble used term 'apparent' velocities' and FLRW metric provided him the 'exact solution' of literal meaning of 'apparent' velocities and therefore chapter was closed in favor of the Big Bang Cosmology. Whereas we have seen already the extent to which the solution was 'exact'. We are forced to conclude that the Big Bang Model is deprived of experimental basis altogether. The only foundation it has is the dubious 'FLRW' metric which is incorrect representative of its own component parts of Friedmann (F) and Lemaître (L). Secondly the whole authenticity of 'FLRW' metric has come from an incorrect claim that Friedmann and Lemaître; or might be Lemaître at least, had derived Hubble type redshift-distance relationship from GR equations before the actual observational discovery of this relationship. The factual position being that the Lemaître had found Hubble type redshift-distance relationship from observational data coupled with the method of distance derivation provided by Hubble himself thus the 'FLRW' metric no more remains 'authoritative' in any sense. The Big Bang Model is not only without experimental basis, it is also without mathematical authority or even philosophical support. By and large, it is only supported by misunderstandings.

* * *

II. OBSERVATIONAL SUPPORT

Big Bang Theory is not based on any empirical data or fact. It is only based on a dubious mathematical 'FLRW' metric. It is regarded as a scientific theory on account of the argument that a number of observed facts are best explained within the framework of Big Bang Model. The second chapter is therefore devoted to analyze this claim of the Big Bang Cosmology. While analyzing each category of claim, we also shall provide alternative explanation only to show that Big Bang based explanation may not always be the best one. However, this book shall remain confined to only the most fundamental claims of the Big Bang Cosmology which are (i) Cosmological Redshift and; (ii) CMBR.

II.I. Cosmological Redshift

If it were Doppler's Effect going on in the light coming from far off galaxies, then it would be right to say that the Big Bang Theory might be a good explanation for the same. But Big Bang Model claims itself to be the best explanation even for cosmological redshift. We accept that Doppler's Redshift is a physical proof of recessional velocity of an object. But Cosmological Redshift, being Redshift-Distance relationship, in isolation, should not be regarded as proof of recessional velocity.

In fact, we do not even regard Doppler's Redshift, in isolation, as proof of recessional velocity of objects. We have first measured Doppler's redshift of various objects and then also have measured physical velocities of those objects and then have formulated a general rule that asserts a positive relationship between recessional velocity and Doppler's redshift.

Now from what sort of general observations have we formulated the same rule for the case of Cosmological Redshifts? Redshift-Distance relationship is not unique in our surrounding real life observations. There is redshift-distance relationship in surface water waves as well as common air waves. Both these observed examples do not lead us towards finding of recessional velocity of source of waves. Then why do Big Bang Cosmologists tell that observed redshift-distance relationship in the light coming from far off galaxies is best explained only within the framework of Big Bang Model which requires that source of light must be having certain recessional velocity?

Mere fact that far off galaxies are redshifted in a way of direct redshift-distance relationship is not the proof that those galaxies are also receding from us at certain 'apparent' velocities. First of all there should be a direct proof to this effect. If a galaxy located at 8 billion light years away is receding away from us at enormous speed then eventually it will reach to the distance of let's say 8.5 billion light years distance. At new location, it will be located at greater distance and therefore now it will be more redshifted than before. Now galaxies are moving away from us at enormous 'apparent' velocities and we are having 100 years old redshift data of many galaxies. Have we noted change in redshift value in any single galaxy so far? The simple reply is no. The lame excuse offered by Big Bang Cosmologists is that galactic distances are so huge that only one hundred year is meaningless in terms of actual distance covered that could have any effect on redshift value. Following is written in previously referred published paper of Indiana University:

"In the cosmological setting which we believe is accurately described by general relativity, we have none of these luxuries! Astronomers cannot wait millions of years to measure quasar proper motions. They cannot, like Highway Patrol officers, bounce radar beams off distant galaxies to establish their relative distances or speeds. Unlike all other forms of motion that have been previously observed, cosmological 'motion' cannot be directly observed. It can only be INFERRED from observations of the cosmological redshift, which general relativity then TELLS US means that the universe is expanding."

Here I agree that there is no observational support to Big Bang Model as cosmological motion cannot be directly measured. But I have objection that 'expansion of universe' is inferred by general relativity. Accurate position is that FLRW metric, being an incorrect representation of Friedmann and Lemaître, has inaccurately described the expansion of universe. Furthermore, expansion ideas of both Friedmann and Lemaître are not fully traceable to actual general relativity.

Although I am not a fan of general relativity – the reasons I shall explain in a separate book; but I acknowledge that general relativity is not guilty of providing the lead towards this faulty and misleading theory of Big Bang. In 1922, Einstein did not accept the idea of expanding universe when Friedmann presented a possibility for the same; he also did not accept idea of expanding universe in 1927 when Lemaître published a case for expanding universe. Einstein did not even accept idea of expanding universe when in 1929, Hubble announced direct relationship between redshift and distance of extra-galactic nebulae. Einstein was only deceived into believing the idea of expanding universe through the manipulated translation of Lemaître's article that was published in year 1931. And it was year 1931 when he changed his mind about Lemaître and Friedmann by abandoning his cosmological constant. A paper about Friedmann mentions following:

"In 1931 Einstein recognized Friedmann's achievement and suggested that his old nemises, the cosmological constant, be expunged from GR."

It is clear for why he did not change his mind in year 1929 when Hubble had announced observational proof of redshift-distance relationship. It was so because Hubble himself had not related his finding with idea of expanding universe. The manipulated translation of Lemaître's article in 1931 was perplexing even for Einstein. If we read only translation, Lemaître had derived that relationship out of GR equations only and the fact was confirmed by Hubble only two years later. That was complete apparent victory of Lemaître and even Einsten had to surrender before magic of mathematics.

Anyhow, the Big Bang Model is devoid of observational support with respect to cosmological redshift. And surprisingly this is the single biggest so called 'observed phenomenon' which is said to be 'best explained by the Big Bang Model'. We have seen so far that neither this observed phenomenon has been linked with similar observed examples of redshift-distance relationship nor any direct proof of explanation in terms of recessional velocities is possible. What has been done by the Big Bang Model is that this observed phenomenon of redshift-distance relationship has been fallaciously linked with other observed examples of redshift-velocity relationship. The term 'redshift' has erroneously become synonym with 'velocity'. It is true that Doppler's redshift does indicate recessional velocity but it does not mean that every kind of redshift indicate recessional velocity. But Big Bang Model has 'best explained' the observed phenomenon of cosmological redshift in exact this erroneous way.

The exact careful statement of Edwin Hubble is that there is direct relationship between 'apparent' velocities and distances of far off galaxies. Since by the term 'apparent' he means 'indeterminate', so let us here take the original statement in precise terms of 'redshift' only. Now the original statement of Edwin Hubble becomes as follows and we assign No.1 to this statement and by rephrasing definition of Cosmological Redshift from an online academic source[29], we get statement No.2:

"There is direct relationship between redshift and distances of far off galaxies." .... (1)

"There is direct relationship between redshift and (expanding) distances of far off galaxies."…. (2)

The definition of Cosmological Redshift as given on stated source is as under:

"In cosmological redshift, the wavelength at which the radiation is originally emitted is lengthened as it travels through (expanding) space."

Now it is only the matter to note the difference between statements 1 and 2. This is what Big Bang Model has actually done with observed facts. In the name of providing 'best available explanation of observed phenomena', this model has distorted actual observed facts by adding unnecessary and unsubstantiated brackets within simple statements of observed facts. It can be argued that bracket is an explanation part which is explaining that redshift is due to (expanding) distance. But my response is that who should accept that redshift is due to (expanding) distance when statement 1 is saying that redshift is due to distance itself?

When there is redshift-speed relationship as in Doppler's Effect, then speed itself is the reason of redshift. It is not right to say that there has to be increasing (recessional) speed so that we may have redshift. Likewise for the case of redshift-distance relationship, distance itself is the reason of redshift. Bracket of (expanding) distance is totally unnecessary to explain redshift-distance relationship. The root cause of this (expanding) bracket was only 'FLRW' metric which was having no real authority to overrule actual observation or override valid reason as we have seen in previous chapter that how in medieval or even primitive style this 'FLRW' metric' had acquired extraordinary authority out of a manipulated translation which depicted as if a hard fact yet to be found through observations was already derived from (metrical) equations of mathematics.

II.I.I. Alternative Explanation of Cosmological Redshift

Various alternatives have been suggested like tired light, gravitational redshifting etc. But standard model has rejected all the already proposed alternatives. Here I am not going to defend those already proposed alternatives. Secondly, it is also not possible to provide direct evidence in support of the alternative proposal. But the suggestion is justified because it is being presented in a philosophy book and not in a science journal who unduly publish Big Bang metaphysical stuff in the name of science. The suggestion will meet the justice for a philosophy book by providing satisfactory argumentative proof. My proposed alternative explanation will also have one extra advantage over Big Bang explanation that unlike Big Bang Model, observed patterns of redshift-distance relationship will be considered rather than considering unrelated redshift-velocity relationship pattern. Given this, the explanation presented here shall become the best available explanation of cosmological redshifts.

II.I.I.I. Examples of Redshift-Distance Relationship in Nature

Actually wavelength-distance relationship is common in our routine daily life observations. Surface water waves have wavelength (redshift)-distance relationship with respect to starting point. Sound waves or common air waves also follow this pattern. We cannot say that in these cases, source of waves was receding away. It is only through faulty 'FLRW' metric that we say that in a similar instance, source of waves was receding away.

II.I.I.II. Why Standard Model Assumes that Light Could Not Follow Simple Natural Pattern?

Surface water waves and sound or air waves travel through medium. But it is assumed that there is no medium of passage for light. Due to this reason, there is no mechanism whereby wavelength of light in space may get increased with distance. Tired Light model was proposed to provide such a mechanism but that model has been discarded on general consensus. In my own opinion, the actual reason of cosmological redshifts is neither tired light nor gravitational redshifting. One way to look at this issue was to see CMBR as ocean where we live. Light travels through this ocean and thus light travels through this medium of CMBR. However this idea might be totally misleading or baseless or it is also possible that there might be truth in this idea but I am not able enough to explore this idea into further depths. But within the assumption of no medium for light, in my humble opinion, the right mechanism that can rightfully explain cosmological redshifts can be found in Huygens Principle. In case CMBR is ocean and serves as medium, then both ocean and Huygens Principle along with minor fractional components of tired light and gravitational redshifting are responsible for observed cosmological redshift. If there is no medium at all, then Huygens principle alone in greater proportion along with tired light and gravitational redshifting in minor proportions account for the observed cosmological redshifts.

II.I.I.III. Huygens Principle – the underlying reason of Cosmological Redshifts

When I was in search of alternative explanation of cosmological redshifts, I conducted a thought experiment. I supposed a source of light; let's say a spherical galaxy, located very far away. I thought if two straight lines of light were originated from spherical surface, then there should be an angle between those two lines. No matter how much small that angle could be, after a sufficient distance, those two lines should get separated from one another. It means that after that sufficient distance, an observer standing between those two lines must not be able to see that galaxy. But another observer whose location might be far behind first one could be able to see the galaxy on account of his position being along one of the lines.

It was not a thought experiment as such but was a simple setting and I wanted answer to the question regarding how first person successfully sees that galaxy when he should not be able to see that galaxy within the framework of these simple settings? I was in search of answer to this question. Eventually, a learned person Mr. Anagh Deshpande[30] guided me to read Huygens Principle as the answer to my question was contained in that principle. Following were his comments which compelled me to consult Huygens Principle:

"That's not how it works! You need to look up Huygen's principle for this, which states that each point on a wavefront is a source of wavelets."

The Wikipedia article[31] describes Huygens Principle in following words:

"In 1678, Huygens proposed that every point to which a luminous disturbance reaches becomes a source of a spherical wave; the sum of these secondary waves determines the form of the wave at any subsequent time. He assumed that the secondary waves travelled only in the "forward" direction and it is not explained in the theory why this is the case."

Now the fact emerging is that with a 'particle line' setup, light is not able to cover every point of the surrounding area of source of light which is simply due to spherical symmetry of straight lines emerging from single point. Actually same was the problem before Huygens in seventeenth century and this fact is the justification how could he manage to present a wave theory of light at such an early time in history. His task was to figure out how light was reaching to each and every surrounding point of source of light. While he presented a sophisticated theory of waves, he actually did not start with waves. He might have started with 'particle lines' of light emitting from spherical point. Now imagine a sphere emitting lines of light like needles. With this setup, in-between two heads of needles there had to be blind spots where from you cannot see the original sphere. In other words, light should not be able to reach many blind spots if it consists of particle lines. To overcome this difficulty, Huygens came up with this Huygens Principle.

So far it is clear that 'particle line' structure of light has been ruled out. Light should propagate in the form of wave. Even waves are curvy and cannot carry original light to the entire outward expanding sphere which is not essentially a sphere but like an infinite cube. Somehow light itself will have to expand; otherwise it will not be able to reach each and every point of far off areas. Huygens identified or conceived that each wavefront must be a source of a new wavelet. Now at least some quarters of modern physics[32] recognize this aspect as a deficiency of this principle because it will lead to the scenario of 'light emitting light' as explained in an article in following words.

"For example, Melvin Schwartz wrote that to consider each point on a wavefront as a new source of radiation, and to add the radiation from all the new sources together, "makes no sense at all", since (he argues) "light does not emit light; only accelerating charges emit light".

Now question is what if this principle is right? Remember that this principle is not fully discarded by modern science. Off and on modern science takes help from this principle. This principle has successfully explained double slit experiments with the exception of only low intensity ds experiments whose failure might be related to some other dynamics of low intensity. So there is no way to altogether reject this principle either on experimental basis or on argumentative grounds. The above article also states following:

"The connection with Huygens' original statement about secondary wavelets is that each wavelet - with the same speed as the original wave - represents a tiny light cone at that point, and Huygens ' Principle asserts that light is confined to those light cones."

Here what is this light cone area? Well, it is the same in-between needle heads area. And why Huygens is saying that wavelet remains confined to cone area? Obviously he is covering all the coming blind spots. He is providing a setup that will ensure that light originating from a point source will reach to each and every point of infinitely scattered sphere.

So with this setup, we are having light emitting light. It should have same implication of absorption and re-emission and thus we will have red-shift which will not only be noticeable over large distances, the redshift will be increased for increased distance as well. In other words, it will be a proper cosmological redshift exhibiting true redshift-distance relationship. It means that light itself cannot cover far off distances without red-shifting and 'expanding universe' is only an illusion.

Somehow light is emitting light and expected dark gaps are being filled. If there is no Huygens Principle, then there is no redshift rather there should be abundance of dark 'cone areas'. The 'cone areas' are being lighten up at the cost of wavelength loss and the overall effect is 'cosmological redshift' which also gives the illusion of 'expanding universe'.

II.I.I.IV. Alternative Explanation of the 'Accelerating Rate of Expanding Universe'

The topic 'accelerating expansion' of universe is more complex than one can expect. Within the accepted meaning of 'Hubble Constant'; officially discovered in year 1929 and whose standard interpretation is all about linear relation between distance of galaxy and 'radial velocity', the galaxies are already 'accelerating' away from one another. The more the distance the more is the radial velocity and the same phenomena of 'increasing velocity' is called 'acceleration'. Within simple and accepted meanings of Hubble law it was therefore already known that galaxies were 'accelerating' away from one another. But we are told that 'accelerating expansion' of the universe was discovered in year 1998. Following is relevant information in a Wikipedia article[33]:

> "The accelerated expansion was discovered in 1998, by two independent projects, the Supernova Cosmology Project and the High-Z Supernova Search Team, which both used distant type Ia supernovae to measure the acceleration. The idea was that these type Ia supernovae all have almost the same intrinsic brightness (a standard candle). Since objects that are further away appear dimmer, we can use the observed brightness of these supernovae to measure the distance to them. The distance can then be compared to the supernovae's cosmological redshift, which measures how fast the supernovae are receding from us. The unexpected result was that the universe seems to be expanding at an accelerating rate. Cosmologists at the time expected that the expansion would be decelerating due to the gravitational attraction of the matter in the universe. Three members of these two groups have subsequently been awarded Nobel Prizes for their discovery. Confirmatory evidence has been found in baryon acoustic oscillations and in analyses of the clustering of galaxies."

Now if we ask the meaning of 'accelerating expansion' of universe by pointing out that galaxies were already believed to be accelerating away from one another, we are told by the experts that meaning of 'accelerating expansion' is that 'rate of expansion' is 'increasing' which means that 'expansion rate' itself is 'accelerating'.

Contrariwise what we actually find is a more complex scenario. The rate of expansion in a direct sense means the value of Hubble Constant. An interesting fact here is that value of Hubble Constant was considered to be 558 Km/sec/MPC during 1930's but now it has been 'corrected' to the figure of only 71Km/sec/MPC. Meanwhile if rate of expansion of universe is on the increase over time then it means that for the older times like galaxies located at distance of 5 billion light years, Hubble Constant should be like 71 Km/sec/MPC (lower rate for older times) and for the recent times like up to the distance of 1 billion light years, the value of 'constant' should be like 558 KM/sec/MPC. But reality check confirms an exactly opposite situation. It turns out that higher value of Hubble Constant of 558 KM/sec/MPC as existed in late 1930's was not due to the fact that Edwin Hubble had observed only nearer galaxies of less than 1 billion light year distance. The answer to the question regarding how Hubble Constant has drastically changed from 558 KM/sec/MPC to only 71 Km/sec/MPC within few decades is not the fact that Edwin Hubble had noted redshift and distance measurements of only nearer 'more accelerated' galaxies located at distance of less than 1 billion light year. Scientists have actually only 'corrected' the value of Hubble Constant to the currently accepted value of 71 Km/sec/MPC from previous inaccurate value of 558 Km/sec/MPC.

If rate of expansion is on the increase then we should expect an increasing trend in value of Hubble Constant which experts now call Hubble 'Parameter' due to its changeability over time. Following graph however shows negative trend evolution of Hubble 'parameter' over time:

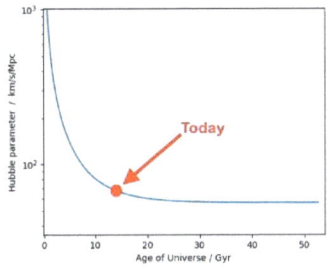

Above graph is showing that with standard model start of time around 13 billion years ago, the value of Hubble Parameter was almost 1000 Km/sec/MPC whereas today it is only 71 Km/sec/MPC. The down slop curve is however flattening which means that the value shall never drop to zero. From the reference of starting point of so called Big Bang, the expansion is in fact decelerating with the provision that due to forever stay of positive though small value of Hubble Parameter, the accelerated receding away of galaxies will keep going on. The expected flattening of Hubble Parameter towards a positive value indicates that gravity will not be able to drop rate of expansion towards decelerating zone and the galaxies shall keep on accelerating away from one another though at a reduced rate than before. From the point of view of an observer standing somewhere in Milky Way, the rate of expansion 'appears' to be 'accelerating' in a proper sense. When observed from earth or nearby space based telescopes, the far off galaxies appear to be receding away at a rate which is greater than the known value of Hubble Constant. The uptrend continues if we observe further deep and far off areas of sky. Scientists concluded that 'Hubble Constant' should better be called 'Hubble Parameter' and this parameter has increasing trend if we look far back in time. We actually only look back in time and what we see is increasing trend for the value of Hubble Parameter. Not only from the reference point of the Milky Way, actually if we look outside while standing on any other galaxy, we would be looking back in time and the far off galaxies would be appearing to be receding away at an 'accelerating rate'.

The actual meaning of 'accelerating expansion' is only that there is uptrend of redshift values at increasing rate and this is from the reference point of the observer. 'Expansion' is only an interpretation part of redshifts and we have seen in the previous section that Huygens' Principle serves as a better explanation of redshifts. We notice redshifts in the light coming from far off sources because mechanism of propagation of light has to ensure that 'cone areas' remain filled with light of the original source. If there is no mechanism in place then there will be no redshift; however there will be dark cone areas. For the light coming from still farther places, the greater cone areas have been filled and thus value of redshift is also high. The relation of distance with the size of cone area is not linear – it is more than linear because with greater distance the angle of cone areas of every upfront wavelet would become wider than before and the overall effect is noticeable only for very long distances just like simple linear relation is also noticeable only after very long initial distance. Universe is not expanding and there is no question of 'accelerated expansion' either. It is 'increasing effort' of Huygens' Principle which is facing additional troubles at greater distances and the extra efforts are being exerted to somehow fill up the cone areas that are now wider than before. Nevertheless this 'increasing redshift at increasing rate' will not continue forever. For the extreme distances, intensity of source light considerably lowers down and it is experimentally known that Huygens' Principle stops functioning or at least starts malfunctioning at low intensities[34]. After that point, consistent and stable image of light source is blurred away; further redshifting is halted and the original light slowly converts into distortion. We receive and detect this extreme redshifted and distorted light of infinite number of galaxies that exist beyond the so called 'observable universe'; wrongfully label it as evidence for the faulty Big Bang Theory and call it 'CMBR', the next section will explain this point in details.

II.II. Cosmic Microwave Background Radiation (CMBR)

CMBR is often projected as the most important observed phenomenon which is 'best explained' by the Big Bang Model. However Big Bang Model itself depends on idea or notion of 'expanding universe' and not on the existence of CMBR. Big Bang Model existed even in that time when CMBR had not yet discovered. Big Bang Model actually takes credit of discovery of CMBR in a complicated way. The following statement of Wikipedia article on CMBR[35] highlights its importance for the Big Bang Model:

> "The discovery of CMB is landmark evidence of the Big Bang origin of the universe."

This Wikipedia article starts with completely inaccurate and misleading description of CMB (or CMBR) in following words:

> "The cosmic microwave background (CMB) is electromagnetic radiation left over from an early stage of the universe in Big Bang cosmology. In older literature, the CMB is also variously known as cosmic microwave background radiation (CMBR) or "relic radiation". The CMB is a faint cosmic background radiation filling all space that is an important source of data on the early universe because it is the oldest electromagnetic radiation in the universe, dating to the epoch of recombination."

Leaving aside the alignment of this description with Big Bang Model, it is factually incorrect because it is depicting CMB as a universal static pond of radiation which is a leftover of original event and now exists everywhere in uniform stationary shape. However, this mistake of the article is corrected under the 'Features' heading in following words:

> "The cosmic microwave background radiation is an emission of uniform, black body thermal energy coming from all parts of the sky."

Therefore now CMB is no more stationary as it is 'coming from everywhere'. Still there is confusion to be got cleared before further proceedings. The common word used for CMB is 'radiation' but the word used here is 'black body thermal energy'. Therefore before moving on, we should clarify this term as well. In simple words, a body is said to emit black body radiation if it emits the same amount of radiation as it receives so that its total temperature remains constant. Following important points from Wikipedia article titled 'Black-Body Radiation'[36] are also worth mentioning:

> "The thermal radiation spontaneously emitted by many ordinary objects can be approximated as black-body radiation. A perfectly insulated enclosure that is in thermal equilibrium internally contains black-body radiation and will emit it through a hole made in its wall, provided the hole is small enough to have negligible effect upon the equilibrium."

According to same Wikipedia article, black-body radiation is also called thermal radiation. CMB is also a kind of thermal radiation which lies mostly in microwaves spectrum range of electromagnetic radiation. Being part of microwave spectrum of electromagnetic radiation, CMB is therefore a kind of infrared invisible light which is coming from everywhere in almost same proportions.

Therefore, according to the Big Bang Model, we are receiving 'remnants' of Big Bang from everywhere. Theory is that Big Bang was like a closed container of point infinite mass. That container started to expand for unknown reasons. The reason is unknown but it is precisely 'known'[37] that within first 10^{-36} to 10^{-32} second, universe (a closed container) had expanded from zero to considerable extent. I do not know this size and neither it is important for our purpose. The important thing so far is however that our not so big container is not emitting light. It has expanded from zero to considerable size at speed greater than speed of light and this anomaly is conveniently justified by saying that light did not exist by that time. Anyhow,

380,000 years after Big Bang, this container universe had reached to the size of 43 million light years[38] and now first time this universe was able to emit light.

So far we have seen the theory of early universe part, now we proceed to see how Big Bang Model became able to predict existence of CMBR?

Big Bang is an expanding universe model and we have reached to an expanded stage of 43 million light years diameter universe which has started to emit light. In fact we have a big glowing container and all the universe is contained inside this universe. Perhaps, 'glow' is also inner-oriented only. The size of this container is 43 million light years. Now, according to Big Bang Model, the 'space' inside this container should keep on expanding. And the original dense light also should have been expanded up to microwave spectrum zone with temperature close to absolute zero. With this type of reasoning and calculations, Big Bang Cosmologists proposed that this type of radiation should universally exist and after few years other researchers accidently found a kind of universally prevailing radiation whose specifications apparently or coincidently almost matched with those proposed by the Big Bang Cosmologists, details thereof can be seen in Wikipedia article and other official sources of information.

First thing is that whole explanation of CMBR under standard model depends on the idea of expanding universe. My question is that what if the universe is not expanding? What is their justification or theory for cosmic microwave background radiation if universe is not in fact expanding? Their CMBR theory fits only under an expanding universe model whereas we already have seen that expanding universe is nothing more than an illusion. But anyhow, we shall examine the case for CMBR within the framework of an expanding universe as well. We have seen that we had a closed container universe of 43 million light years diameter just 380,000 after Big Bang. Our own galaxy, in whatever form, existed somewhere within this container. For the sake of simplicity, we can assume that our galaxy existed at the center of that container and also emitted the thick hot glow at that time. The container universe as a whole then kept on expanding at speed greater than speed of light. The thick light emitted by our galaxy at that time also kept on expanding (in wavelength due to expansion of space) but it was travelling at speed of light because according to standard model, light always travels at speed of light.

Keeping in view the total age of universe of 13.8 billion years under standard model, approximately 13.799 billion years have been passed since the first thick light was emitted by our galaxy. That light is now exactly 13.799 billion light years away from us. But radius of total universe has reached to 45 billion light years. At the distance of 13.799 billion light years from the original location of our galaxy, now there is another galaxy named 'Agronexa'. Agronexa galaxy has received the light of (newly born) Milky Way galaxy in expanded form in microwave spectrum zone of electromagnetic radiation and people of Agronexa galaxy may also call it CMBR.

Likewise, the first CMBR that we received was the expanded light of newly born Agronexa galaxy. With this scheme, although CMBR is the first ever light emitted by the universe but the first CMBR that we received might not have come from edge of the universe. In fact, there can always be galaxies located at distance beyond the source of CMBR that we are actually receiving.

Under simplified settings where universe is not expanding at speed greater than speed of light, then we can easily point out anomalies in this justification of CMBR. Source of CMBR is said to be state of universe as it was some 380,000 years after Big Bang. But just after 200 or 600 million years of Big Bang we are having visible galaxies. Therefore, the source of CMBR stayed for a maximum period of 600 million years only. It means that radiation emitted only during this short period is now visible on Earth.

For example if intelligent Dinosaurs or some other intelligent animals lived 600 million years ago and reached to the theory of Big Bang then they could not have (accidentally) found this CMBR proof of Big Bang. It is now we humans have found these radiations but the radiations will vanish in maximum period of 600 million years for now. It is like a great coincidence that we live in exact this era where we can find this radiation. After all it (CMBR) is light and coming from all directions at speed of light. Its visibility is only for that duration for which the source remained in existence. Being normal or close to normal light, this is not like echo of Big Bang that should stay forever.

But the above results are under our simplified settings. Standard Model defenders would argue that we were contained inside the source of CMBR that was 43 million light years in diameter. It was expanding by way of 'metrical expansion'. Original light emitted within one portion of the source was trying to reach other portions while those other portions were moving away at enormous speeds that were greater than speed of light, by way of metrical expansion of space. In this way, all portions will keep on receiving CMBR of all other portions for indefinite time. The crucial thing is expansion of universe speed greater than speed of light. Here we choose to not raise preliminary objection that if expansion speed is greater than speed of light then simply we should not see CMBR. But in another scenario if light speed is also expanding metrically then universe is not expanding at all with respect to speed of light. Actually within the framework of an expanding universe, there arises a question that if universe can expand at speed greater than speed of light (i.e. superluminal speed) then why light itself cannot move at same superluminal speed within same metrically expanding universe?

If space is expanding metrically due to which physical galaxies are receding away at superluminal speeds then light also should travel on same expanding space and thus should travel at same superluminal speed. With these settings, within the framework of an expanding universe we have a universe which is static with respect to speed of light at least. Within such an expanding universe where speed of expansion is not faster than speed of light – more precisely, where superluminal expansion of universe is counterbalanced by superluminal speed of light, there also the maximum period of visibility of CMBR should have been the actual infancy period of only 600 million years because after this period, galaxies had formed according to standard model. But under standard model, the visibility period of CMBR shall stay for indefinite – like infinite period because expansion of universe speed is superluminal but speed of light itself is not superluminal. But case for fixed velocity of light in an expanding universe is not as simple in standard model as well. In a paper titled "Superluminal Recessional Velocities"[39] by Tamara M. Davis and Chales H. Lineweaver (University of New South Wales), a case for superluminal velocities of light has been presented. Following are few quotes out of this paper:

"Here we show that galaxies with recession velocities faster than the speed of light are observable and that in all viable cosmological models, galaxies above a redshift of three are receding superluminally."

"How can photons reach us from regions of space that are receding superluminally? How can they cross from the light grey into the dark grey? In Figure 2 the boundary between the light and dark grey regions is the Hubble sphere, the distance, DHS(t) = c/H (t), at which galaxies are receding at the speed of light. The comoving distance to the Hubble sphere increases when the universe decelerates and decreases when the universe accelerates. The Hubble sphere is not an horizon of any kind; it passes over particles and photons in both directions."

Argument of this paper is that cosmological redshift is not the subject of special relativity (SR) but is the subject of general relativity (GR) therefore infinite redshift does not imply velocity of light equal to c. This paper asserts that visible galaxies having redshift values three or above are receding superluminally and thus light is also reaching us superluminally.

Though I am content with the results but I do not agree with the reasoning and thus also do not fully agree with the results. This paper breaks at least one taboo however that light cannot travel at superluminal speed even in a metrically expanding universe. My objection on reasoning of this paper is that Cosmological Redshift has nothing to do with general relativity at all. Abstract of this paper starts with incorrect information that Hubble's Law, v = HD (recession velocity is proportional to distance), is a theoretical result derived from the Friedmann-Robertson-Walker metric. Fact is that Hubble's law is not a theoretical result. It was first time derived by Lemaître in his 1927 French article where he derived approximately same rule on the basis of observational data of redshifts and method of derivation of distance provided by Hubble. Hubble's Law might not be first presented by Hubble himself but it had never been earlier derived on the basis of any method other than Hubble's method. General Relativity had no role in the derivation of Hubble's law prior to presentation of same law by Hubble himself. But unfortunately, general relativity acquired a deceptive role in the derivation of Hubble's law after 1931 when Lemaître published a manipulated English translation of his 1927 French article. He manipulated translation in such a way that the translated article would show as if Hubble Law type relationship was derived from GR equations and the actual relation with Hubble's method of finding distance was omitted in the translation. Therefore, general relativity has no real relationship with expanding universe. But it is true that within the framework of a superluminally expanding universe, Light also has to move superluminally. Here I explain it with simple example.

Metric Expansion is actually like 'Big Bang' that is happening and creating 'space' everywhere all the time. Let's say 1mm distance is becoming 2mm over one billion years, 2 mm distance becoming 4 mm in same one billion year and so on. This is an over-exaggerated example, but in standard model, almost 13 billion light years is a distance which is known as Hubble sphere such that beyond which it is considered that galaxies are moving away at superluminal speed and therefore cannot be seen. But whatever lies within Hubble sphere is visible. Now suppose we are observing a galaxy 'A' close to the boundary of Hubble sphere. That galaxy 'A' is already receiving light from those galaxies that are beyond of our Hubble sphere. Now suppose there are large broadcasting reflectors installed on galaxy 'A' such that they can send received images of beyond galaxies to us. In this way, we can actually see what is beyond our own Hubble sphere. Here nothing can bar galaxy 'A' from receiving light of beyond galaxies and then nothing can bar us from receiving light of galaxy 'A' which includes reflected light of beyond galaxies. Tamara M. Davis' paper also accepts that light can move at superluminal speed. But if superluminal expansion of universe is exactly counterbalanced by the superluminal speed of light then we have an expanding universe framework which is static at least with respect to speed of light. Assuming there is no alternative explanation of cosmological redshift then with this 'static' universe, I may raise doubt whether there be any cosmological redshift exists or not. But one thing is clear that period of visibility of CMBR cannot be more than the period of physical existence of source. Under standard model, CMBR is not an extra redshifted normal light but is light of something which existed prior to the existence of galaxies. That something existed only for 600 million years and with a universe static with respect to speed of light, the period of visibility of that something also should be same 600 million years.

II.II.I. Alternative Explanation of the 'CMBR'

We have seen that standard model has explanation of CMBR within framework of only an expanding universe with fixed speed of light and where universe is expanding superluminally. With any other setup, CMBR will have either improper or no explanation at all. Here the standard model has a doubtful mathematical model of expanding universe but the observed 'proof' (CMBR) rightfully depends on notion of expanding universe. This is like fool proof explanation of CMBR within the framework of metrically expanding universe. But a possibility of superluminal speed of light itself in an expanding universe is a loophole within this otherwise fool proof explanation. Secondly, CMBR can be acknowledged as an observed proof for an expanding universe only if such a radiation could not be possible to occur other than in an expanding universe. Mere fact that Big Bang Cosmologists had the 'prediction' of the existence of similar kind of radiation is insufficient proof in support of expanding universe theory given that this kind of radiation could also be possible in a non-expanding universe. But this possibility was denied and radiation was projected as possible to occur only under the Big Bang Model thus presented to be rightfully serving as 'observational proof' to the Expanding Model. The main supporting point was the uniformity of the CMBR across whole of skies. It was argued that the source must be the whole of the universe in initial form as it existed 380,000 years after big bang. Then Universe expanded and so did the radiation which we are now uniformly receiving from all the directions.

Here if we could show that uniform receipt of radiation from all the directions is possible in a non-expanding universe, then whole case of CMBR as proof of Big Bang Model should collapse at once. But before showing that radiation can be received from all the directions, it is also important to show that radiation emitted by many ordinary objects can be approximated as black-body radiation. Following is relevant quote from Wikipedia article titled 'Black-Body Radiation':[40]

"The thermal radiation spontaneously emitted by many ordinary objects can be approximated as black-body radiation. A perfectly insulated enclosure that is in thermal equilibrium internally contains black-body radiation and will emit it through a hole made in its wall, provided the hole is small enough to have negligible effect upon the equilibrium."

Now it is time to show that under standard model itself, it is possible to uniformly receive radiation from all parts of the sky. This is famously known as 'Olbers Paradox' and in most simple words, Wikipedia article[41] has to say following about it:

> "The darkness of the night sky is one of the pieces of evidence for a dynamic universe, such as the Big Bang model. In the hypothetical case that the universe is static, homogeneous at a large scale, and populated by an infinite number of stars, then any line of sight from Earth must end at the (very bright) surface of a star and hence the night sky should be completely illuminated and very bright. This contradicts the observed darkness and non-uniformity of the night."

We see that this 'Olbers Paradox' serves as an evidence for Big Bang Model which apparently tells us about a finite but expanding universe. On the other hand, a 'hypothetical' infinite universe which is populated by infinite number of homogeneously distributed stars should be characterized by a completely illuminated and bright night sky. Since night sky is dark, so 'hypothesis' of infinite universe having infinite number of stars is incorrect and therefore the same fact also serves as evidence for Big Bang Model.

There are certain confusions associated with Olbers Paradox that we should resolve first. First thing is that when Heinrich Wilhelm Olbers (1758–1840) postulated this paradox, at that time there was concept of perfect homogeneous distribution of infinite stars in 'hypothetical' infinite universe as there was no concept of separate 'island universes' of galaxies having huge voids in between. But the paradox, in principle, holds even now because galaxies are uniformly distributed and should exist with uniform distribution infinitely if the hypothesis of infinite universe is true.

The second confusion arises because this paradox is also called 'dark night sky paradox'. This title of the paradox is misleading because outer space is always dark even in presence of sun. The bright daylight at earth is not directly due to sunlight but is due to glow acquired by atmosphere of earth by sunlight. At daytime, radiation coming from sun is sufficient enough that our atmosphere acquires bright glow. At nighttime, radiation coming from stars is not sufficient enough that our atmosphere could acquire bright glow and thus our night remains dark. Therefore, Olbers Paradox is accurate at face value and dark night sky should mean finite size of universe with due support for the Big Bang Model.

Now we come to see the other side of the picture. Initially the Big Bang Model did imply a finite model of universe as the universe started from a point and expanded with a finite speed for a finite duration of time. But later on data gathered through better space telescopes compelled cosmologists to reinterpret their beloved 'FLRW' metric. Following are words of a famous internet Physics writer Mr. Victor T. Toth[42] on this topic:

> "There is no evidence that the universe is finite. The simplest model (a so-called Friedmann-Lemaitre-Robertson-Walker universe) that fits the data actually shows a "flat", infinite universe.
> But even if the universe is "closed", which implies finite, it does not have an edge. Topologically, it's the same idea as a circle that is finite but without endpoints. Or the surface of a sphere that is finite but without boundary. The universe is not one-dimensional (like the circle) or two-dimensional (like the surface of a sphere) but three-dimensional, but its presumed finiteness is in the same spirit, so to speak."

Here Victor T. Toth, while being affirmative on infinite universe also does not fully rule out possibility of finite or closed universe. But anyhow, available data is telling that there is no evidence for a finite universe. Problem with Big Bang Cosmologists is that they always start telling even about observational data by first telling us about 'FLRW' metric which was neither relevant in past nor has anything to do with latest observational data. Simple fact is that available data is suggesting that there is no evidence for a finite universe. Big Bang Cosmologists only have to complete their sentences by using 'FLRW' metric in any mode or form. We can understand that they now reinterpret their 'FLRW' metric to somehow accommodate infinite universe idea within Big Bang Cosmology. In fact, under footnote No.11 of his 1922 paper, Friedmann has accepted that time from creation till now could be infinite and thus radius of universe also could be infinite.

Therefore now other picture is clear. There is acceptable possibility of an infinite universe under standard model. Only thing required is that Olbers Paradox should not be discussed while discussing possibility of infinite universe under standard model i.e. the exact same approach adopted by Victor T. Toth. However, within other contexts, Olbers Paradox may be continued to be used as a 'proof' for Big Bang Model. By now, readers should have well realized the type of this standard model whether it is really a form of science or what kind of a manipulation it is. Perhaps manipulations are unavoidable because standard model is subject to internal contradictions whose one example is the possibility of infinite universe and at the same time total denial of infinite universe using Olbers Paradox.

The identified problem with the idea of an infinite universe is that if it is true then our night sky should be bright. But exact this 'problem' is the clear indication that at least, in principle, it is possible to receive uniform radiation from all parts of the sky in a non-expanding infinite universe having homogeneous distribution of infinite many stars. And we are actually receiving uniform radiation from all parts of the sky. Off course Olbers, in 19th century did not know about Hubble's law which states that light coming from farther distance should be more redshifted. Unfortunately this law has been interpreted only in expanding universe context whereas literally this is relation only between redshift and distance of source of light. We should forget here about relationship of velocity with redshift because it is a separate relationship and comes from Doppler's findings and not from Hubble's law. Keeping in view the fact that light coming from far off sources has to be considerably redshifted and by observing the type of radiation that we actually receive uniformly from all parts of the sky, it is now easy to conclude that CMBR is the redshifted (to microwave zone of spectrum) light of very far off galaxies that exist in our infinite universe which is homogeneously populated by infinite many galaxies. Olbers had not actually presented a 'paradox'. He pointed out a possibility – that if read properly with Hubble's law and properly interpreted in terms of Huygens Principle, there was a rightful 'prediction' of the existence of CMBR in 19th century. Huygens Principle does not let light to travel far off distances without redshifting and the underlying purpose is that light may reach to every point surrounding the source of light. After travelling considerable distance, light gets redshifted and distorted too much that Huygens Principle no more contributes to the availability of solid consistent image of source of light. When light has travelled a sufficient distance then at the receiving end a low intensity light has reached and it is experimentally known that Huygens Principle fails to properly function within low intensities of light.[43] After the point when Huygens Principle starts malfunctioning due to low intensities of received light, further redshifting might be stopped but distortion continues. What seems right is that CMBR is this distorted light of very far off galaxies which we cannot read or figure out to estimate the kind of source of origin. What we receive is distorted light which is redshifted to the zone of microwave portion of spectrum of electromagnetic radiation. We call it CMBR and this is proof that our universe is far larger than the size that can actually be calculated under BIG Bang Cosmology. Perhaps this CMBR is the proof that we live in an infinite sized universe.

If we accept that expansion is an illusion, even then there are certain anomalies that apparently go in favor of finite sized universe having finite age and Big Bang model also exploits those anomalies. We are told that scientists can find only less than average signals of higher than Hydrogen atoms in very remote (and thus early) galaxies. Then there is also range of distance where signals of higher than Helium atom are also rare. Big Bang Cosmologists present these 'facts' as proof of their model. Actual fact is that we receive not only redshifted but also distorted light from very far off galaxies. Due to the fact that light received is distorted, the source of origin cannot be rightfully traced, at least by using expansion based methodology. This is the reason that scientists cannot see full signals of higher than Helium elements in far off distance and see only marginal traces of higher than Hydrogen atoms at still farther distance i.e. more than 13 billion light years. For the still farther distances, light is fully distorted and redshifted to invisible microwaves portion thus nothing is 'seen' except for an invisible uniform brightness of CBMR in microwaves portion of spectrum of electromagnetic radiation.

Future of the Big Bang Model is that after realizing mounting evidence against this model coming from 'non-authoritative' sources, science authorities will try to find excuse against Big Bang Cosmology from within standard model. This book also proposes them the same route as a safe escape strategy.

* * *

¹ https://en.wikipedia.org/wiki/Edwin_Hubble *(In 1929, in his first published paper, Hubble examined the relation between distance and redshift of galaxies. Combining his measurements of galaxy distances with measurements of the redshifts of the galaxies by Vesto Slipher, and by his assistant Milton L. Humason, he found a roughly linear relation between the distances of the galaxies and their redshifts,[8] a discovery that later became known as Hubble's law.*

This meant, the greater the distance between any two galaxies, the greater their relative speed of separation.)

² https://en.wikipedia.org/wiki/Edwin_Hubble *(In reality, Georges Lemaître, a Belgian Catholic priest and physicist, predicted on theoretical grounds based on Einstein's equations for General Relativity the redshift-distance relation two years before the discovery of Hubble's law. However, many cosmologists and astronomers (including Hubble himself) failed to recognize the work of Lemaître, with, to date, no remaining papers or verification that they found or accepted any link between Lemaître's work and Hubble's measurements.)*

³ Sandage, Allan (1989), "Edwin Hubble 1889–1953", *The Journal of the Royal Astronomical Society of Canada*, Vol. 83, No.6. Retrieved March 26, 2010. *(Hubble believed that his count data gave a more reasonable result concerning spatial curvature if the redshift correction was made assuming no recession. To the very end of his writings he maintained this position, favouring (or at the very least keeping open) the model where no true expansion exists, and therefore that the redshift "represents a hitherto unrecognized principle of nature.)*

⁴

http://www.physics.umd.edu/grt/taj/675e/Luminet_on_Lemaitre_history.pdf

⁵ https://en.wikipedia.org/wiki/Edwin_Hubble *(In 1931 he (Edwin Hubble) wrote a letter to the Dutch cosmologist Willem de Sitter expressing his opinion on the theoretical interpretation of the redshift-distance relation:[33]*

Mr. Humason and I are both deeply sensible of your gracious appreciation of the

papers on velocities and distances of nebulae. We use the term 'apparent' velocities to emphasize the empirical features of the correlation. The interpretation, we feel, should be left to you and the very few others who are competent to discuss the matter with authority.)

[6] https://en.wikipedia.org/wiki/Friedmann%E2%80%93Lema%C3%AEtre%E2%80%93Robertson%E2%80%93Walker_metric (*In 1935 Robertson and Walker rigorously proved that the FLRW metric is the only one on a spacetime that is spatially homogeneous and isotropic.*)

[7] https://academic.oup.com/mnras/article/91/5/483/985165

[8] https://arxiv.org/pdf/1305.6470.pdf

[9] https://arxiv.org/pdf/1305.6470.pdf

[10] https://arxiv.org/pdf/1305.6470.pdf

[11] http://imgsrc.hubblesite.org/hvi/uploads/science_paper/file_attachment/69/pdf.pdf

[12] He did not actually abandon cosmological constant. Only thing is that he adopted value assigned to it by Friedmann.

[13] https://en.wikipedia.org/wiki/Arthur_Eddington

[14] https://en.wikipedia.org/wiki/Subrahmanyan_Chandrasekhar

[15] http://einsteinpapers.press.princeton.edu/vol6-trans/433

[16] https://map.gsfc.nasa.gov/universe/bb_tests_exp.html

[17] https://en.wikipedia.org/wiki/Cosmological_constant

[18] https://en.wikipedia.org/wiki/Alexander_Friedmann (This dynamic cosmological model of general relativity would come to form the standard for both the Big Bang and Steady State theories. Friedmann's work supports both theories equally, so it was not

until the detection of the cosmic microwave background radiation that the Steady State theory was abandoned in favor of the current favorite Big Bang paradigm.)

[19] https://en.wikipedia.org/wiki/Steady_State_theory

[20] *(Alexander Friedmann and the origins of modern cosmology:* Ari Belenkiy - Phys. Today 65(10), 38 (2012); doi: 10.1063/PT.3.1750, http://dx.doi.org/10.1063/PT.3.1750

[21] "On the Curvature of Space" Aleksandr Friedmann – Translation by Brian Doyle.

[22] De-Sitter's model is now-a-days projected as expanding universe model and even inflationary period is also derived from de-Sitter's present (modified) model. By the time of Friedmann, it was a stationary model of Universe. Edwin Hubble, after 1929 had requested de-Sitter to provide explanation of redshifts. Hubble himself remained skeptical to the idea of expansion but de-Sitter adopted expansion ideas after 1929.

[23] https://en.wikipedia.org/wiki/Hubble%27s_law

[24] https://en.wikipedia.org/wiki/Metric_expansion_of_space

[25] http://www.pnas.org/content/15/3/168.full

[26] https://en.wikipedia.org/wiki/Vesto_Slipher

[27] https://en.wikipedia.org/wiki/Edwin_Hubble

[28] http://cecelia.physics.indiana.edu/life/redshift.html

[29] http://astronomy.swin.edu.au/cosmos/c/cosmological+redshift

[30] The question asked by me was --- https://www.quora.com/How-can-the-light-of-a-galaxy-reach-to-each-and-every-corner-of-surrounding-universe-at-a-distance-of-a-few-billion-LYs ---- The comments of Mr. Anagh

Deshpande can be seen under the comments section of answer by Mr. Jess. H. Brewer.

[31] https://en.wikipedia.org/wiki/Huygens%E2%80%93Fresnel_principle

[32] http://www.mathpages.com/home/kmath242/kmath242.htm

[33] https://en.wikipedia.org/wiki/Accelerating_expansion_of_the_universe

[34] https://en.wikipedia.org/wiki/Huygens%E2%80%93Fresnel_principle

[35] https://en.wikipedia.org/wiki/Cosmic_microwave_background

[36] https://en.wikipedia.org/wiki/Black-body_radiation

[37] https://en.wikipedia.org/wiki/Inflation_(cosmology)

[38] https://www.quora.com/What-was-the-size-of-the-universe-380-000-years-after-the-Big-Bang

[39] https://arxiv.org/pdf/astro-ph/0011070v2.pdf

[40] https://en.wikipedia.org/wiki/Black-body_radiation

[41] https://en.wikipedia.org/wiki/Olbers%27_paradox

[42] https://www.quora.com/What-is-beyond-the-edge-of-the-universe-1/answer/Viktor-T-Toth-1

[43] https://en.wikipedia.org/wiki/Huygens%E2%80%93Fresnel_principle

www.ingramcontent.com/pod-product-compliance
Lightning Source LLC
Chambersburg PA
CBHW040229220526
45473CB00001B/179